The No-Tofu
Vegetarian
Cookbook

The No-Tofu

Vegetarian

Cookbook

Sharon Sassaman Claessens

HPBooks

HPBooks
Published by The Berkley Publishing Group
200 Madison Avenue
New York, NY 10016

First edition: July 1997

Published simultaneously in Canada.

The Putnam Berkley World Wide Web site address is
http://www.berkley.com

Library of Congress Cataloging-in-Publication Data

Claessens, Sharon.
    The no-tofu vegetarian cookbook / Sharon Sassaman
Claessens.—1st ed.
    p.   cm.
    Includes index.
    ISBN 1-55788-269-X
    1. Vegetarian cookery.   I. Title.
TX837.C4975   1997   96-52652
641.5'636—dc21   CIP

Printed in the United States of America

10   9   8   7   6   5   4   3   2   1

Notice: The information printed in this book is true and complete
to the best of our knowledge. All recommendations are made
without any guarantees on the part of the author or the publisher.
The author and publisher disclaim all liability in connection with
the use of this information.

To Adam and Anne,
who offer encouragement and
give love and support

Contents

Acknowledgments

Many, many thanks to friends and family who have taste-tested myriad dishes and offered approval, praise or constructive feedback as warranted. My gratitude to Terry Scott, who patiently transcribes my preferred longhand to computer, is deep and sincere. I am also fortunate to have an experienced, well-educated editor (with a background in nutrition) to review, revise and correct, with grace and care, my manuscripts: Jeanette Egan brings many gifts to this editorial task, and I am grateful to work with her. Again, I thank the staff at The Berkley Publishing Group for their support and the care and effort they put into producing a quality product at a fair price to the consumer. For anyone who contributed in any way to the completion of this book, I say "Thank you."

Introduction

Why a No-Tofu Cookbook?

No food has become more synonymous with vegetarian cooking over the years than tofu. Used for centuries in Oriental cuisines, tofu (soybean curd), a bland-tasting, high-protein food, has made its inexorable way across most of the vegetarian landscape, invading a variety of ethnic cuisines in which it had previously never existed. Tofu was traditionally found in recipes for Chinese, Korean and Japanese main dishes and soups. But now recipes feature tofu in casseroles, quiches, pizzas, salads and salad dressings, sandwiches, soufflés and dips. It's been scrambled for breakfast, stuffed into everything from avocados to zucchini, and even sweetened and blended to create "cheesecake," "puddings" and chocolate "mousse."

Nearly every vegetarian cookbook currently being sold features tofu. In fact, a newcomer exploring vegetarian cooking might, and often does, ask: "Will I have to eat tofu?" The answer, as you may have suspected, is: "No." There are hundreds of wonderful vegetarian dishes that provide ample protein, vitamins, minerals, fiber and essential trace nutrients without tofu.

Though you may have heard all this before, tofu is a wonderful food. It is low in fat. It is convenient to use. It can be flavored in any variety of ways. But if you've picked up this book, these arguments already may be lost on you. Perhaps you haven't even tasted tofu, but you've seen it in your supermarket swimming in liquid in its little plastic box or pressed into firm, spongy-looking cakes, and decided: "Not for me!" Well, someday you may change your mind. But for now, let's explore the world of vegetarian eating without tofu.

About Vegetarianism

In addition to the cuisines of the Orient, vegetarian dishes have been a traditional part of many of the world's great menus. The populations of the Middle East, Mexico, Africa and other parts of the world may have eaten meat, chicken or fish on special occasions, or whenever it was available. But much of their diet relied on grains, beans, nuts, vegetables and other fruits of the earth . . . in short, vegetarian cuisine.

Because they may not regularly have eaten meat, these ancient peoples developed a dietary system that actually made optimum use of the proteins and other elements available in their simple foods. Unknown to them, these vegetable foods also contained something we now call essential amino acids. (For those of you who like the nitty-gritty details, they are: lysine, valine, leucine, isoleucine, histidine, methionine, threonine, tryptophan and a real tongue-twister, phenylalanine.)

Balancing Protein

Interestingly enough, these nine amino acids don't appear in the same proportions in all foods. Vegetable foods have "incomplete" protein, which means that some of the necessary amino acids are either missing or in too short supply. Even more interestingly, traditional cultures began combining vegetable foods that complemented each other— one compensating for lack in the other, and vice versa. What this did was to maximize the usefulness of all those amino acids by yielding "complete" protein, the kind that we know today is found in a steak or chicken breast. And these "primitive" people figured all this out without so much as a test tube!

Legumes such as dried beans and lentils "complement" grains such as rice, wheat, barley and corn. (An example: Traditional Mexican bean tacos are beans plus corn tortillas.) Seeds complement beans (i.e., the Middle Eastern staple known as hummus tahini, a pureed dip and spread made with sesame seeds and chickpeas).

When Frances Moore Lappé wrote *Diet for a Small Planet* in the late 1960s, a great deal of attention was focused on the necessity for eating complementary proteins at every meal when eating a meatless diet. Since then, further research into vegetarian diets has resulted in a more relaxed attitude. No longer must each dish be planned out for maximum protein benefit.

First, we know that most of us get far more protein from our diets than necessary. And excess protein, over and above what the body can use, is eventually stored . . . as fat. Second, we find that complementary foods do not have to be eaten at the same time. In fact, their protein-boosting power seems to kick in when foods are eaten anytime during the day. Third, we find that most foods, including many vegetables and fruits as well as the more commonly accepted beans and grains, contain some protein. That means if we vary our diet with a wide selection of foods, it is almost certain we will consume sufficient protein.

Reasons for Eating a Vegetarian Diet

Many people are attracted to vegetarian meals because they have been shown to be more healthful. Non–meat eaters exhibit lower rates of heart disease and cancer, two diseases that increasingly plague our richly abundant Western society.

Some are motivated to become vegetarians by their concern for the environment and lack of sufficient food for a significant portion of the world's population. They know, for example, that it takes sixteen pounds of soybeans to create just one pound of beef. (Significantly, soybeans are the only vegetable food that already contains complete protein of a quality equal to that found in meat.) Others have religious reasons, or object to the unnecessary slaughter of animals for the table.

One of the advantages of a vegetarian diet, even one that includes low-fat dairy products and eggs, seems to be an easier control of weight. Because meat products can be so high in fat, it is difficult, unless one has a *very* active life-style, to burn up all those calories, and a very active life-style is not something many of us have these days.

For example, my ancestry is Pennsylvania Dutch. My grandparents were farmers who enjoyed meals that were plain and simple, but rich in the meats and poultry that they raised. A lot has changed since the farm was sold to enlarge a budding college campus. When my forebears got up from their large dinner (served in the middle of the day, followed by a lighter "supper"), they pruned orchards, tilled, sowed and harvested the soil, milked cows, thrashed grain and produced huge kettles of apple butter by hand.

Now, the copper apple butter kettle, once blackened with use, is handsomely lacquered and holds logs (cut by strangers) beside my wood-burning stove. My apple butter comes in a jar. My only effort, occasionally, is getting the lid off. Instead of wrestling a plowshare through open fields, I navigate my pen across paper (having not yet succumbed to a computer). And my son, rather than tending livestock, pitching hay and cleaning the barn as my father did, pours out a cup of kibble, then takes a walk with the puppy he is raising as a 4-H project for The Seeing Eye.

Evenings, too, have changed. On the farm, shelling peas, cutting and drying apple *schnitz*, mending and quilting kept hands busy. These days, as we gather for our favorite evening show around the television set, our hands are more likely busy transporting snacks.

With activity levels so vastly different, we must adjust our diets, too. The beef roasts, pork, sausage and gravies that graced my grandparents' table have given way to lighter fare. Now, more often than not, the dinner meal is meatless. A constant has been the selection of grains, herbs, vegetables, fruits and dried beans my grandparents had raised or bartered for.

My grandparents probably would not have understood the concept of vegetarian meals, though when times were lean, they certainly had meals without meat. But that was considered a time of deprivation. They might have been astounded at the concept of meatless meals by choice—and certainly by the idea of meatless meals as satisfying, delicious and even festive.

My grandparents may not have appreciated that mushrooms, broccoli, beans, seeds, nuts, cornmeal, oats, wheat or rice could be high in protein. Or that these foods, when eaten in combination, could yield a protein as nourishing as that of any meat. And, unless we overdo on oils, high-fat dairy items or nut butters, it is actually difficult to get too much fat.

Types of Vegetarians

Vegetarians who avoid all animal foods (including eggs, dairy foods, gelatin and even honey) are called "vegans." (Vegans often take B_{12} supplements to avoid anemia. Manufactured by beneficial bacteria, B_{12} is also found in animal products.) Vegetarians who eat eggs are "ovo-vegetarians," and those who include dairy products are "lacto-vegetarians." Vegetarians who eat eggs and dairy foods are "lacto-ovo-vegetarians."

Some people who occasionally eat fish and/or chicken but refrain from eating red meat still consider themselves vegetarians.

Planning Vegetarian Meals

How do we assemble wholesome, balanced vegetarian meals? By piecing together a variety of foods into delicious main dishes, such as those found in this book, we can create square meals from vegetarian sources. Like a quilt, these meals may have a different pattern than the one we're accustomed to. If, for example, your typical evening meal has consisted of meat, two vegetables and a salad, a vegetarian dinner will appear slightly different. Main dishes will often contain the vegetables we are used to seeing "on the side."

These recipes contain simple, colorful and nutritious foods. Menus can be assembled from dishes made from foods that you and your family will enjoy. By selecting a variety of dishes throughout the week, and avoiding junk foods, you will be supplying yourself with the protein, fiber and other nutrients needed for good health.

If you are not already a vegetarian, you may want to start gradually, including one or two vegetarian meals per week. Or serve much smaller amounts of meats with a small serving of a vegetarian main dish, gradually eliminating the meat altogether, as you become more acquainted with the dishes you like. You may be surprised, as you switch to more vegetarian meals, at the energy and "lightness" you feel after eating.

Whichever way you choose to use this book, I hope you experience new pleasures in eating and build a stronger basis for good health.

Appetizers & Soups

Easy Appetizers

Versatility and ease are two bywords of this section. For versatility, look to such dips as Olive-Herb Hummus Tahini and Roasted Sesame Eggplant Dip. Either protein-rich dip can be served in a bowl or spread on a pita bread half as the basis for a sandwich. Add thinly sliced vegetables and you're set!

For ease, the microwave is often used to reduce cooking time. Another time-saver is the use of canned dried beans instead of home-cooked beans.

Super Soups

Not only can soups be a meal, but they are among the most nutritionally dense, low-calorie foods you can eat. In fact, soups have been shown to be effective in weight loss. One reason might be that in eating a hot soup, we are forced to slow down. This gives our "appestat" time to say "Just enough, thanks!" *before* we've overeaten.

Soups are also good for a sense of well-being. Nothing smells quite as reassuring as a pot of soup simmering on the stove. Add a good bread and a salad, and your meal is complete.

Garlic-Lovers' Salsa Cruda

If you aren't crazy about garlic, reduce the amount to 4 cloves. Otherwise, just bid farewell to all your friends (unless they're garlic-lovers, too!). Serve with low-fat tortilla chips.

1 large onion, finely minced
6 to 7 cloves garlic (about half
    an entire head), peeled and
    minced
2 teaspoons olive oil
6 to 8 tomatillos, finely chopped
1 tablespoon ground cumin

1/2 tablespoon ground coriander
1 (28-oz.) can whole tomatoes,
    drained and chopped
1 (8-oz.) can tomato sauce
2 tablespoons red wine vinegar
2 tablespoons minced cilantro

Place onion and garlic in a large microwave-safe bowl and toss with oil. Microwave on HIGH 2 minutes. Add tomatillos, cumin and coriander and microwave on HIGH 1 minute. Stir in tomatoes, tomato sauce, vinegar and cilantro.

Makes about 4 cups, about 16 servings.

Education

Salsa can be an important feature of a vegetarian diet based on beans. For while dried beans provide important amounts of iron, it is not as readily available to the body as the iron found in meats. Vitamin C, which is present in abundance in salsa, helps make iron from vegetable sources more available. So saucing up your Mexican-style dishes will give you more than just flavor.

Green Onion Guacamole

As an appetizer dip or side dish with Mexican-style dishes, guacamole packs a punch with vitamins A and C. The avocado should "give" slightly, to be most flavorful.

1 large, ripe avocado
2 green onions, minced
1/4 cup bottled mild salsa
2 teaspoons fresh lemon juice

1/2 garlic clove, minced
1/4 teaspoon salt
1/8 teaspoon freshly ground black pepper

Peel avocado and remove pit. Mash in a medium bowl with a stainless steel fork. Stir in green onions, salsa, lemon juice, garlic, salt and pepper. Serve with low-fat taco chips or vegetables.

Makes about 1 1/2 cups, about 10 servings.

Education

Although avocados are high in fat (about 75 percent of their calories), they are very nutritious and can be included in a sensible diet occasionally. To put it in perspective, a little over half of an avocado has less fat than an equal weight in lean ground beef, less than 1 cup of potato salad with mayonnaise or a green salad with 2 tablespoons of regular Italian dressing.

Sesame Roasted Eggplant Dip

Roasting the eggplant and the garlic provides a rich flavor to this appetizer. Serve with pita wedges or melba toast.

1 fresh garlic bulb (about
    16 cloves)
1/4 cup extra-virgin olive oil
1 eggplant (about 1 lb.), cut into
    1-inch-thick slices
1 onion, chopped

1 cup chopped mushrooms
1/2 cup fresh lemon juice
1/4 cup sesame tahini
1/2 teaspoon salt
1/8 teaspoon freshly ground
    black pepper

Preheat oven to 450F (230C). Cut the papery tip from the top of the garlic, just down to the tops of the garlic cloves. Place garlic in a small microwave-safe dish. Drizzle with 1 teaspoon of the oil. Microwave on HIGH 1 minute.

Arrange eggplant slices in one layer on a nonstick baking sheet. Lightly brush eggplant slices on both sides with oil. Wrap precooked garlic in foil and place on baking sheet. Roast in oven 20 minutes, or until eggplant is soft. Set aside to cool.

Heat the remaining oil in a medium skillet over medium heat. Add onion and cook, stirring frequently, until onion is translucent, 3 to 4 minutes. Add mushrooms and cook, stirring occasionally, until mushrooms are tender and their liquid has evaporated, about 5 minutes.

Remove and discard skin from eggplant slices. Coarsely chop eggplant and place in a food processor. Squeeze baked garlic into food processor, discarding any hard cloves. Add onion and mushroom mixture, lemon juice, tahini, salt and pepper and process until nearly smooth. Spoon into a bowl and chill several hours or overnight.

Makes about 2 cups, 16 servings.

White Bean Spread with Capers & Sage

It may sound fancy, but it's an easy appetizer to make. As with other spreads and dips on these pages, it is versatile, and can also be used as a sandwich spread. For a sandwich, fill pita pockets with spread and a mixture of thinly sliced vegetables and some diced tomato or lettuce.

1 slice soft, whole-wheat bread,
    crusts removed
2 tablespoons water
1 2/3 cups cooked (or 15-oz. can)
    white kidney beans, drained
2 tablespoons chopped walnuts
2 tablespoons fresh lemon juice
1 tablespoon olive oil

3 garlic cloves, minced
1 tablespoon minced capers
2 teaspoons minced fresh sage
    leaves or oregano leaves or
    1/4 teaspoon dried
1/4 teaspoon salt
Dash of ground cayenne

Crumble bread into a small dish and sprinkle with water. Set aside to soak. In a food processor, combine beans, walnuts, lemon juice, olive oil and garlic. Process until almost smooth.

Add bread and process until smooth. Stir in capers, herbs, salt and cayenne. Spoon into a bowl and chill several hours or overnight to blend flavors. Serve with pita wedges, melba toast or raw vegetables.

Makes about 2 cups, 16 servings.

Education

Your own cooked beans will be firmer and have a fresher flavor than canned beans. But in dishes such as this, when the beans will be pureed, the convenience of canned beans outweighs the disadvantages. To cook beans from scratch, see page 91.

Olive-Herb Hummus Tahini

So delicious, guests at a tasting party hovered around this dish. Serve with an assortment of raw vegetables or pita wedges.

1 2/3 cups cooked (or 15-oz. can) chickpeas, drained
1/4 cup sesame tahini (see Note, below)
1/4 cup fresh lemon juice
1/3 cup water
2 garlic cloves, minced
1/4 cup fresh parsley
1 teaspoon fresh oregano leaves

1/4 teaspoon salt
1/8 teaspoon freshly ground black pepper
1/4 cup minced, pitted Greek olives
1/2 teaspoon olive oil, for garnish
Dash of ground cayenne or paprika, for garnish

Combine the chickpeas, tahini, lemon juice, water and garlic in a food processor and process until smooth. Add the parsley, oregano, salt and pepper and process until combined. Add olives and process just until combined. Cover and chill 1 hour or overnight. Scoop mixture into a serving bowl and allow to come to room temperature. Drizzle with oil and sprinkle with cayenne or paprika. Serve with raw vegetables or pita triangles.

Makes about 2 cups, 16 servings.

Note Sesame tahini, a traditional Middle Eastern ingredient, is a paste made up of ground sesame seeds. When using tahini, stir well to incorporate any oil that has separated during storage. Store opened containers in the refrigerator.

Romaine Lettuce Leaves with Vegetable Puree

Choose the small, crisp inner leaves of romaine. This filling for the leaves is easy, but elegant, too. Spoon it into the leaves or go all the way and pipe it in with a pastry bag and fancy tip. The filling can also be served as a dip with crackers.

1 large head romaine
1 tablespoon olive oil
2 cups chopped leeks (white part
    only)
1 cup water
1/3 cup red lentils
1/4 cup finely grated carrots
1 teaspoon minced fresh herbs
    (parsley and tarragon, basil
    or marjoram)

1/4 teaspoon salt
1/8 teaspoon freshly ground
    pepper
Dash of freshly grated nutmeg
Fresh herb sprigs, for garnish

Remove and set aside large romaine leaves for salad. Pull apart inner leaves and rinse. Shake excess water and place in a plastic bag with a damp paper towel. Refrigerate until ready to use.

Heat oil in a large nonstick skillet over medium-low heat. Add leeks and cook, stirring occasionally, until nearly tender, about 10 minutes. Add water, lentils, carrots, herbs, salt, pepper and nutmeg. Bring to a boil, reduce heat and simmer, uncovered, until the lentils are tender and the liquid is nearly absorbed, about 30 minutes.

Place leek mixture in a food processor or blender and process until smooth. Transfer to a bowl, cover and refrigerate until chilled. Spoon into romaine leaves. Garnish with herb sprigs.

Makes about 1 1/2 cups, 12 servings.

Savory Olive-Tomato Garlic Toast

An Italian-style appetizer features nonfat garlic bread topped with an olive mixture and colorful fresh tomatoes. Everything can be made a day ahead, then assembled and heated the day of serving.

1 to 2 loaves French or long,
    narrow Italian bread
14 garlic cloves
1 cup pitted ripe olives
2 teaspoons balsamic vinegar
1 teaspoon capers, drained
2 teaspoons olive oil
1 red tomato, peeled, seeded and
    chopped
1 yellow (or another red) tomato,
    peeled, seeded and chopped

1/4 cup thinly sliced green onion tops
1 tablespoon thinly sliced fresh
    basil leaves or 1/2 teaspoon
    dried basil
1 teaspoon minced fresh oregano
    leaves or 1/4 teaspoon dried
    oregano
1 teaspoon minced fresh parsley
1/8 teaspoon salt
1/8 teaspoon freshly ground black
    pepper

Preheat oven to 425F (220C). Cut bread on the diagonal into 24 (1/2-inch-thick) slices. Peel and halve 12 garlic cloves. Run the cut side of each garlic half over both sides of each piece of bread. Place the bread slices in one layer on baking sheets. Bake 5 minutes, turning once, or until crisp and lightly toasted. Allow to cool.

Peel and coarsely chop the remaining 2 garlic cloves. Combine chopped garlic, olives, vinegar, capers and oil in a blender or food processor. Process until nearly smooth, stopping to scrape down the sides of the container as necessary. Set aside.

Combine chopped tomatoes, green onions, basil, oregano, parsley, salt and pepper in a small bowl.

To assemble, spread each toasted bread slice with some of the olive mixture. Divide tomato mixture among the bread slices. Arrange slices on baking sheets and bake 2 to 3 minutes, or until the toppings are heated through.

Makes 24 slices, about 12 servings.

Ginger-Tahini Dip

Many dips contain dairy products such as cream cheese. Here, however, protein-rich sesame tahini provides a flavorful counterpoint for crisp vegetables.

1/3 cup sesame tahini	1 garlic clove, minced
1/4 cup fresh lemon juice	Carrot sticks
1/4 cup water	Broccoli flowerets
1 tablespoon ginger juice (see Note, below)	Cauliflowerets
1 tablespoon teriyaki sauce	Green onions
1 teaspoon honey	Fennel bulb, sliced

Combine tahini, lemon juice, water, ginger juice, teriyaki sauce, honey and garlic in a food processor or blender and process until smooth. All the dip ingredients can be placed in a serving bowl and stirred until combined. At first the mixture will appear curdled, but keep stirring and it will become creamy. Serve with assorted raw vegetables.

Makes about 1 cup dip.

Note Grate fresh gingerroot and squeeze to yield juice. Depending upon freshness of the root, 1/4 to 1/3 cup of grated gingerroot will yield 1 tablespoon juice.

Easy Pickled Vegetables

Add a tasty twist to raw vegetables by marinating them overnight. Rice vinegar is available reasonably priced at Oriental food markets.

3/4 cup rice vinegar	1 large garlic clove, halved
1 tablespoon canola oil	1 small cucumber, halved and
1 teaspoon salt	seeded
1 teaspoon teriyaki sauce	2 cups carrot sticks
1 teaspoon honey	1 cup cauliflowerets

In a wide-mouth quart jar, combine the vinegar, oil, salt, teriyaki sauce and honey. Shake until combined. Add garlic.

Cut the cucumber into thin half-rounds. Add cucumber, carrots and cauliflower to marinade. Seal jar and shake. Place in refrigerator and allow to marinate at least overnight, shaking a time or two. Remove and discard garlic. Remove vegetables from marinade and serve with wooden picks.

Makes 10 servings.

Note   You can reuse the marinade for up to two weeks. Try adding green onions, broccoli, radishes and other vegetables.

Vegetable Broth

The heartiness of this broth comes from cooking the onions to a brown flavorful concentration. Using your own delicious stock will be convenient in the long run, giving slow-simmered flavor to even the quickest soups.

2 tablespoons canola oil
1 large onion, chopped
1 fennel bulb with top, coarsely
    chopped
1 large carrot, diced
1 small turnip, diced
1/4 cup diced mushrooms or
    1 tomato, seeded and chopped

3 parsley sprigs
3 fresh thyme sprigs
2 whole cloves
1 small bay leaf
1/2 teaspoon salt
5 cups water

Heat the oil in a large pot over medium-low heat. Add onion and slowly cook, stirring often, until the onion is well browned and very tender, about 20 minutes. Add the fennel, carrot, turnip and mushrooms and cook 1 minute, stirring. (Vegetables need not be peeled, as the stock will be strained.) Add herbs, salt and water and bring to a boil over medium-high heat. Reduce heat, cover and simmer about 1 1/2 hours, or until vegetables are very soft.

Strain broth, discarding vegetables. Chill broth. Remove any fat from the top of the stock before using.

Makes 4 cups broth.

Variation

I like the delicate flavor imparted by the mildly anise-flavored fennel. Two large stalks of celery can be substituted for the fennel, if you desire.

Education

Making your own vegetable broth is one more way to save money on food. Not just because you are saving at the supermarket, where most of what you pay for when buying "broth" is water, but because broth improves with older, more mature vegetables.

This means the outer stalks of celery, which may not be visually appealing for salads or main dishes, are the best for broth. Also, vegetable trimmings—carrot peelings, mushroom and tomato skins, etc.—can all lend wonderful flavor to broth. Place a self-sealing bag in the freezer (in the refrigerator if you plan to make broth at least once a week) and collect trimmings and less attractive vegetables for broths.

Slow-Cooker Vegetable Broth

Just made for the long, slow simmering that makes a good broth, a slow cooker can help you create this kitchen basic nearly effortlessly. A quick browning of onions and carrots intensifies the stock's hearty flavor.

1 tablespoon canola oil	1 dried mushroom
1 large onion, chopped	1/4 cup parsley stems and leaves
1 large carrot, diced	3 fresh thyme sprigs
10 cups water	1 bay leaf
3 cups coarsely chopped green	2 whole cloves
leek tops	1 teaspoon salt
1 celery stalk, sliced	1/4 teaspoon freshly ground black
1 turnip, diced	pepper

Heat the oil in a nonstick skillet over medium-high heat. Add the onion and carrot and cook, stirring frequently, until the vegetables begin to brown, 8 to 10 minutes. Place in a slow cooker with water, leek tops, celery, turnip, mushroom and seasonings. Simmer on high 6 to 8 hours, or overnight, until vegetables are very soft. Strain off stock and discard vegetables.

Makes about 8 cups.

Variation

In place of leek tops, use a combination of two or more of the following: loosely packed spinach stems and leaves, lettuce leaves, tomato skins, fresh mushroom stems, or diced parsnips along with some additional chopped celery and carrot to equal 3 cups vegetables.

Quick Vegetable Broths

Not everyone can take the time to make a slow-simmered vegetable stock. Here are several short-cut stocks that can still provide a flavorful base for vegetarian dishes. Other alternatives are canned broth and vegetable broth cubes, diluted in water.

Quick Broth I

In addition to soups, this combination is useful when cooking quick brown rice.

Combine about 2/3 cup (5.5-ounce can) eight-vegetable juice with 1 1/4 cups water. Season with salt and pepper.

Makes 2 cups.

Quick Broth II

Save water from cooking beans (soy water is very tasty) and cooking vegetables (potato water, for example). These broths add nutrients as well as flavor. Strongly flavored vegetables, such as cauliflower, broccoli or cabbage, will yield a distinctive broth that can be used with soups made with more of the same.

Refrigerate or freeze for convenient later use. Ice cube trays can be used for small amounts. Once frozen, the "stock" cubes can be removed and stored in a self-sealing plastic bag.

Dilled French Lentil Stew

A trio of lentils—petite French, red and green (or brown)—yields a fine, tasty, attractive stew when combined with leeks. Europeans enjoy leeks more than we do in America, but we are definitely catching on! Serve this wonderful soup with crusty French bread.

1 tablespoon olive oil
1 cup chopped leeks (white parts
    only)
1/4 cup finely chopped carrot
3 garlic cloves, minced
1/2 cup French lentils
1/2 cup green or brown lentils
1/4 cup red lentils

5 cups water
1 bay leaf
1/2 teaspoon salt
1 cup eight-vegetable juice
1 tablespoon minced fresh dill
    weed
Dash of freshly ground black
    pepper

Heat the oil in a nonstick Dutch oven over medium heat. Add leeks and cook, stirring often, until the leeks are wilted and begin to brown, 4 to 5 minutes. Add carrot and garlic and cook, stirring, 1 minute.

Add lentils, water, bay leaf and salt and bring to a boil. Reduce heat, cover and simmer 35 minutes, or until lentils are tender.

Add juice and dill and simmer 5 minutes. Remove bay leaf and season with pepper.

Makes 4 servings.

Curried Ginger Butternut Stew

Chickpeas give a heartiness to this vegetable-rich stew with its colorful palette of crisp-tender vegetables.

1 tablespoon soy margarine
2 onions, chopped
2 garlic cloves, minced
1 teaspoon finely minced
    gingerroot
1 teaspoon curry powder
1 teaspoon ground coriander
4 cups vegetable broth

1 medium butternut squash
1 (16-oz.) can chickpeas
1 cup frozen yellow whole-kernel
    corn
1/2 cup finely chopped broccoli
    flowerets
1/2 cup finely chopped cauliflowerets
1/4 cup finely diced red bell pepper

Melt the margarine in a nonstick Dutch oven over medium-high heat. Add onions and cook, stirring frequently, until the onions are translucent, 3 to 4 minutes.

Add garlic, ginger, curry and coriander and stir until combined. Add broth and bring to a boil. As it heats, peel and cube squash. Add squash to broth. Return to a boil, then reduce heat, cover and simmer 20 to 30 minutes, or until the squash is tender.

Puree half the soup in a food processor or blender. Return to the Dutch oven and add chickpeas, corn, broccoli, cauliflower and bell pepper. Return to a boil, reduce heat and simmer, uncovered, 4 to 6 minutes, or just until the vegetables are crisp-tender.

Makes 6 servings.

Education

If you like the flavor of ginger, but not the fibrous pieces of the root swimming in your soup, think small. By grating gingerroot and squeezing out the liquid (you'll be surprised how much that dry-looking root can contain), you can get the flavor without the pieces.

30-Minute Minestrone

Everyone's favorite can be ready in half an hour! This vegetarian version recreates the symphony of colors and textures of the traditional soup in half the time.

2 tablespoons olive oil
1 cup chopped onion
1 cup thinly sliced celery
1 cup diced carrots
2 garlic cloves, minced
4 cups vegetable broth
1 cup finely chopped cabbage

1 cup diced zucchini
1 2/3 cups (or 15-oz. can) white
    kidney beans, drained
1 (14 1/2-oz.) can Italian-style
    stewed tomatoes
2 tablespoons minced fresh parsley

Heat the oil in a large, nonstick Dutch oven over medium-high heat. Add the onion, celery and carrots and cook, stirring frequently, until the vegetables begin to brown, 5 to 7 minutes. Add garlic and cook, stirring, 1 minute.

Add broth, cabbage and zucchini and bring to a boil. Reduce heat, cover and simmer 10 minutes, or until vegetables are firm-tender.

Meanwhile, place half the beans in a blender with the liquid from the stewed tomatoes and process until smooth. Dice tomatoes and add to soup. Add blended bean mixture and parsley and return to a boil. Reduce heat, cover and simmer 5 minutes.

Makes 6 to 8 servings.

Variation

Add 1 cup cooked pasta, such as macaroni, to the soup. If you desire, you can sprinkle each serving with some freshly grated Parmesan cheese.

Garlic-Lovers' Vegetable Soup

If you love garlic, you know that simmering tames its wild, pungent flavor, promoting a happy blend with other ingredients.

1 tablespoon olive oil	4 cups vegetable broth
1 cup chopped onion	1/2 teaspoon salt
2 tablespoons minced fresh garlic	1/2 teaspoon dried marjoram
1 celery stalk, thinly sliced	1/4 teaspoon dried thyme
1 large carrot, diced	1/4 teaspoon ground coriander
1 ripe tomato, peeled, seeded and diced	1/2 cup finely chopped broccoli flowerets
1/2 cup diced zucchini	1 tablespoon minced fresh parsley
1/2 cup finely chopped cabbage	1 tablespoon minced fresh basil

Heat the oil in a large, nonstick Dutch oven over medium-high heat. Add the onion and cook, stirring frequently, until onion begins to brown, 5 to 7 minutes. Add garlic and cook, stirring, 1 minute.

Add celery, carrot, tomato, zucchini, cabbage, broth, salt and dried herbs and bring to a boil. Reduce heat, cover and simmer 5 minutes. Add broccoli, parsley and basil and simmer 5 minutes.

Makes 6 servings.

Education

Use garlic to protect your blood. Not only does folklore assure us that garlic can repel vampires, scientists now say that garlic can protect our blood from other evils as well. For example, garlic seems to prevent excess blood clotting. It also seems effective in lowering blood levels of cholesterol and triglycerides, reducing the risk of heart disease.

Gazpacho

Have you ever heard gazpacho described as a salad you eat with a spoon? Try this wonderful chilled summer soup and see if you agree. Soup can be thinned, if desired, with ice-cold water.

2 garlic cloves
1 teaspoon salt
2 to 3 slices white bread, crusts removed
3 tablespoons red wine vinegar
1/4 cup olive oil
1 teaspoon ground cumin
1/2 teaspoon ground coriander
2 1/2 cups tomato juice
3 cups peeled, seeded and finely chopped garden-ripe tomatoes

1 cup finely diced green bell pepper
1/3 cup finely chopped green onions
1/3 cup finely chopped red onion
1 cucumber, peeled, seeded and finely chopped
2 tablespoons minced fresh cilantro
2 tablespoons minced fresh parsley
Dash of red hot pepper sauce
Croutons, for garnish
Cilantro sprigs, for garnish

On a cutting board, mince the garlic and mash with the side of the knife with the salt to make a paste. Tear bread into small pieces to make 1 1/2 cups. In a blender, combine garlic paste, bread, vinegar, oil, cumin, ground coriander and 1 cup of the tomato juice and process until smooth.

In a large bowl, combine bread mixture, the remaining juice, tomatoes, bell pepper, green onions, red onion, cucumber, minced cilantro and parsley and season to taste with hot pepper sauce. Cover and chill 3 hours or more.

Makes 6 servings.

Carrot & Potato Chowder

Save the cooking water from the potatoes; it is part of the base for this light, creamy soup. I like to use a variety of potatoes: yellow, red and russet.

3 large potatoes, peeled and diced
2 cups water
1/2 teaspoon salt
2 tablespoons soy margarine
2 large carrots, diced
1 small celery stalk, thinly sliced
1 small turnip, finely diced

1 large onion, diced
2 shallots, minced
1 cup vegetable broth
1 tablespoon minced fresh parsley
1/2 teaspoon ground coriander
1/4 teaspoon dried marjoram

Bring the potatoes, water and salt to a boil in a large saucepan over medium-high heat. Reduce heat, cover and simmer 15 to 20 minutes, or until the potatoes are tender. Drain potatoes, reserving cooking water. Add a little water, if necessary, to make 2 cups. Process the potatoes in a food processor just until smooth.

While potatoes are simmering, melt the margarine in a large saucepan over medium heat. Add the carrots, celery, turnip, onion and shallots and cook, stirring occasionally, until onion is firm-tender, about 4 to 5 minutes. Add cooking water from potatoes, pureed potatoes, vegetable broth, parsley, coriander and marjoram and bring to a boil. Reduce heat, cover and simmer 10 minutes, or until vegetables are tender.

Makes 4 servings.

Vermont White Bean & Cabbage Soup

Canned beans make this quick and substantial soup easy to prepare.

2 teaspoons olive oil
1 large onion, chopped
1 large garlic clove, minced
1/2 teaspoon ground cumin
1 cup vegetable broth

2/3 cup tomato or eight-vegetable
   juice
2 cups chopped cabbage
1 2/3 cups cooked (or 15-oz. can)
   white beans, drained

Heat the oil in a large, nonstick saucepan over medium-high heat. Add the onion and cook, stirring often, until the onion is translucent, about 3 to 4 minutes. Add the garlic and cumin and cook, stirring, 1 minute.

Add broth, juice, cabbage and beans and return to a boil. Reduce heat, cover and simmer 15 to 20 minutes, or until the cabbage is tender.

Makes 4 servings.

Education

Be it ever so humble, cabbage seems to be one of the most potent cancer-fighting foods. Cabbage (and other members of the *Brassica* genus, such as Brussels sprouts and broccoli) contains indoles, which seem to ward off cancer. In addition to the soup above, red or white cabbage can be shredded and added to nearly any cooked soup or to salads.

Rio Chili Soup

Accompany this hearty soup with cornbread and salad for a complete meal. The texture stays interesting with the firm-tender zucchini.

1 tablespoon olive oil
1 large onion, chopped
1/2 cup diced red bell pepper
1 zucchini, diced
2 garlic cloves, minced
2 teaspoons chili powder
1 teaspoon ground cumin
1 teaspoon ground coriander
1/2 teaspoon ground yellow mustard
1/2 teaspoon dried oregano

1/2 teaspoon salt
1/4 teaspoon freshly ground black pepper
2 cups eight-vegetable juice
2 cups vegetable broth
1 2/3 cups cooked (or 15-oz. can) kidney beans, drained
1/4 cup finely chopped ripe olives
1/4 cup minced fresh parsley

Heat the oil in a large, nonstick Dutch oven over medium-high heat. Add the onion and cook, stirring occasionally, until the onion is translucent, about 3 to 4 minutes. Add the bell pepper, zucchini and garlic and cook, stirring, 2 minutes. Add chili powder, cumin, coriander, mustard, oregano, salt and pepper and stir until combined.

Add juice and broth and bring to a boil. Reduce heat, cover and simmer 10 minutes, or until the zucchini is firm-tender. Add beans, olives and parsley and heat through.

Makes 4 servings.

Variation

Unexpectantly have two more for dinner? Or want enough soup to freeze planned-overs? Add another cup of juice or stock, another can of beans (you could add small red beans or black beans to the kidney beans) and 1 cup whole-kernel corn. Add 1/4 cup of chopped green olives, if you've got them.

Fennel Split-Pea Soup

The mild anise flavor of fennel offers a graceful note to a hearty soup.

2 tablespoons olive oil
2 cups chopped fennel
1 1/2 cups diced carrots
1 1/2 cups chopped onions
3 garlic cloves, minced
1/2 teaspoon ground coriander
1/8 teaspoon ground cloves
1/2 bay leaf

8 cups vegetable broth
4 cups water
1 pound dried green split peas
2 potatoes, peeled and diced
1/2 teaspoon salt
1/8 teaspoon freshly ground black pepper

Heat the oil in a large, nonstick Dutch oven over medium-high heat. Add the fennel, carrots and onions and cook, stirring frequently, until vegetables begin to brown, about 8 to 10 minutes. Add garlic, coriander, cloves and bay leaf and stir until combined.

Add broth, water, split peas, potatoes, salt and pepper and bring to a boil. Reduce heat, cover and simmer 1 hour, stirring occasionally, or until split peas are tender.

Remove bay leaf. Puree the soup in a food processor or blender. Reheat if needed.

Makes 8 servings.

Education

Like any legume, split peas are an excellent source of protein. And unlike some, split peas are relatively quick-cooking and need no presoaking. Eat soup with whole-grain bread or over brown rice if you want to boost the protein even more.

Red Lentil Soup

Carrots help give this soup its wonderful, delicate color. A touch of minced parsley is the only garnish needed.

2 teaspoons olive oil
1 large onion, chopped
1 cup diced carrots
2 garlic cloves, minced
1/4 teaspoon ground coriander
1/4 teaspoon salt

1/3 cup red lentils
1/4 cup quick-cooking barley
4 cups vegetable broth
1 tablespoon minced fresh parsley,
    for garnish

Heat the oil in a large saucepan over medium-high heat. Add the onion and cook, stirring frequently, until tender, about 4 to 5 minutes. Add carrots, garlic, coriander and salt and cook, stirring, 1 minute.

Stir in lentils and barley until combined. Add broth and bring to a boil. Reduce heat, cover and simmer 15 or 20 minutes, or until ingredients are tender.

Place soup, one half at a time, in a blender or food processor and process until smooth. Reheat if necessary. Pour soup into individual bowls and garnish with parsley.

Makes 4 servings.

Education

A slowly simmered split pea soup is wonderful, but when you can't take the time, try red lentils. They cook quickly, and teamed with quick barley, are protein-rich, besides! Soups based on complementary proteins, like beans and barley or lentils and rice, can take center stage for any meal. Serve with whole-grain crackers, biscuits or bread and a fresh salad.

Balsamic Borscht

Fresh beets have such an incomparable flavor that it is worth just that little extra effort to prepare this colorful Russian soup classic from scratch.

2 cups water	2 cups vegetable broth
3 or 4 red beets	1 1/2 cups finely shredded cabbage
2 large carrots	1 tablespoon balsamic vinegar
1 large onion	Minced fresh parsley, for garnish

Place the water in a large saucepan and bring to a boil. Peel and dice beets (there should be about 3 cups), dice carrots and chop onion. Place vegetables in the boiling water and return to a boil. Reduce heat, cover and simmer 20 minutes.

Coarsely puree the vegetables with liquid in a food processor or blender and return to the saucepan. Add broth, cabbage and vinegar and return to a boil. Reduce heat, cover and simmer 15 minutes. Serve hot or chilled, garnished with minced parsley.

Makes 6 servings.

Variation

If you eat dairy foods, top each serving with 1 tablespoon plain yogurt or low-fat sour cream.

Education

If you see red after eating beets, don't be alarmed. Many people are unable to metabolize betacyanin, the pigment that makes beets reddish-purple. This means the pigment passes through their bodies unchanged. Though harmless, the results can be unnerving if you don't realize the cause.

Quick Bean Soup for a Small Crowd

You can please a hungry bunch with this heart-warming soup, based on a can of refried beans!

2 tablespoons olive oil
2 cups chopped onions
3 garlic cloves, minced
1 jalapeño chile (optional), seeded
    and minced
2 teaspoons chili powder
2 teaspoons ground cumin
1 teaspoon ground coriander
1/2 teaspoon dried oregano
1/2 teaspoon ground yellow mustard
1 (16-oz.) can vegetarian refried
    beans

3 cups vegetable broth
2 (14 1/2-oz.) cans Mexican-style
    stewed tomatoes
3 1/3 cups cooked [or 2 (15-oz.) cans]
    red kidney beans, drained
1 2/3 cups cooked (or 15-oz. can)
    black beans, drained
2 flour tortillas, for garnish
1/2 cup minced fresh cilantro, for
    garnish

Heat the oil in a large, nonstick Dutch oven over medium-high heat. Add the onions and cook, stirring frequently, until the onions are crisp-tender and begin to brown, about 5 to 6 minutes. Add the garlic and jalapeño, if using, and cook, stirring frequently, 2 to 3 minutes.

Add chili powder, cumin, coriander, oregano and mustard and cook, stirring, just until combined. Add refried beans and 1 cup of the broth and stir until smooth. Pour in remaining broth and stewed tomatoes, dicing any large pieces of tomato. Add kidney and black beans and bring to a boil. Reduce heat, cover and simmer 5 minutes.

Meanwhile, cut the tortillas into thin strips, about 2 x 1/4 inches. When soup is heated, pour into individual bowls and garnish with tortilla strips and cilantro.

Makes 10 servings.

Leek & Potato Soup

Ever wonder what to do with the lovely bunches of leeks you see in the supermarket? This soup is simplicity itself. You can also serve it cold, but then you'll call it vichyssoise.

3 cups diced peeled russet potatoes
3 cups thinly sliced leeks (white
    part with 1 to 2 inches of pale
    green)

5 1/2 cups vegetable broth or water
1 teaspoon salt

Place ingredients in a large saucepan and bring to a boil over medium-high heat. Reduce heat, cover and simmer 25 to 30 minutes, until vegetables are tender. Puree in a blender or food processor until smooth. Serve hot or chill.

Makes 4 servings.

Variation

Stir 1 tablespoon buttermilk into each serving, or top with a spoonful of low-fat sour cream and sprinkle with minced fresh chives.

Education

Leeks can be an important vegetable for vegetarians because of their significant iron content. Iron is most often associated with meat. While leeks are not as rich in iron as beef, they far surpass chicken, pork and many fish. Per serving, leeks approximate the iron found in lamb.

Leeks are milder-tasting and more nutritious than onions, and can be substituted for onions in most dishes. They must be carefully cleaned, however, to remove the soil and grit that gets lodged in their leaves. Slitting the leaves from the white root end and fanning them open under running lukewarm water will dislodge this grit, or the leek can be cut into 2-inch lengths and the layers separated and rinsed in a bowl of lukewarm water.

Chilled Cantaloupe Soup

So simple, so refreshing and light, this delicately spiced soup could also be served as dessert on a warm evening. Its success depends upon a ripe, juicy and flavorful melon.

1 large, ripe cantaloupe, chilled
Dash of ground cardamom
Dash of ground ginger

4 whole strawberries, for garnish
Mint sprigs, for garnish

Peel and seed melon. Finely dice 1/2 cup of the flesh and set aside. Coarsely cube remaining melon. There should be about 5 cups. Add melon, cardamom and ginger to a blender or food processor and process until smooth.

To serve, pour soup into individual bowls. Garnish with reserved finely diced melon and a strawberry and/or a mint sprig.

Makes 4 servings.

Education

Try to get more melons into your diet. Not only are melons low in calories because of their high water content, they are also extremely nutritious. Melons are a good source of beta carotene, potassium and vitamin C. Oh, and, of course, an excellent source of sweet flavor.

Simple Winter Tomato Soup

When fresh tomatoes aren't in season, this tasty soup can fit the bill.

2 teaspoons olive oil
2 teaspoons soy margarine
1 large onion, finely chopped
1/4 cup finely chopped celery
1/4 cup finely chopped carrot
3 garlic cloves, minced

1/4 teaspoon dried marjoram
1/4 teaspoon dried thyme
1/4 teaspoon salt
1 (28-oz.) can peeled diced tomatoes
2 cups water
Minced fresh parsley, for garnish

Heat the oil and margarine in a nonstick Dutch oven over medium heat. Add the onion, celery, carrot and garlic and cook, stirring frequently, until the vegetables begin to brown, about 5 to 7 minutes.

Add the marjoram, thyme, and salt and cook, stirring, 1 minute. Add tomatoes with juice and water and bring to a boil. Reduce heat, cover and simmer 15 minutes. Serve garnished with parsley.

Makes 4 servings.

Cornucopia Fruit Soup

This soup is so versatile you can pick whatever fruits are in season. (See below for suggestions.) Serve hot or cold as a starter or even for dessert.

4 cups pitted, cored, seeded and,
    if necessary, coarsely chopped
    fruits

3 1/2 to 4 cups water
1 1/2 tablespoons fresh lemon juice
1 to 2 tablespoons sugar

Combine fruits, enough water to just cover, lemon juice and sugar to taste in a large saucepan and bring to a boil over medium-high heat. Reduce heat, cover and simmer 20 minutes, or until fruits are tender.

If you want to include fruit skins in the final dish, puree soup in batches in a blender. Or the soup can be strained through a colander or sieve to remove skins. Serve soup hot or cold.

Makes 6 servings.

Variations

Use one main fruit, such as fresh apricots, with small amounts of other fruits, such as grapes, apples or pears. Combine peaches, plums, apricots and blueberries or cherries. Small amounts of dried fruits, such as raisins, apricots or prunes, may be added. The tastiest fruits will yield the most flavorful soups, of course, but this is a nice way to use up fruits that are slightly overripe, or have become mealy.

Watercress Soup

Watercress is often used in salads, but can easily be made into a quick and tasty soup for two. The bright green color of this soup is outrageously gorgeous.

2 teaspoons soy margarine	1 cup vegetable broth
1/3 cup finely chopped onion	1/4 teaspoon salt
1 shallot, minced	Dash of freshly ground pepper
1 bunch watercress, rinsed and	Thin lemon slices, for garnish
chopped	Watercress leaves, for garnish

Melt the margarine in a medium saucepan. Add the onion and shallot and cook, stirring frequently, until the onion is translucent, about 3 to 4 minutes. Add watercress and stir until wilted. Add broth and bring just to a boil. Remove from heat and puree in a food processor or blender. Serve hot or chilled. Season with salt and pepper. Garnish by floating a lemon slice on the soup, topped with a leaf of watercress.

Makes 2 servings.

Education

Watercress is a member of that illustrious family of cruciferous vegetables believed to be effective in preventing cancer. Since, as a general rule, the darker the green, the more nutritious, one would rightly assume that watercress stands out. In fact, it is high in vitamin C, beta carotene and calcium, as well as folacin and fiber.

Overnight Bean & Barley Soup

A variety of beans, none of which needs presoaking when prepared in a slow cooker, adds up to hearty winter fare. Serve with cornbread or other whole-grain bread and a salad for a satisfying cold-weather meal.

1 cup dried red kidney beans
1 cup dried white kidney beans
1/2 cup dried black turtle beans or
    small red beans
1/2 cup quick-cooking barley
1 large onion, chopped
1 tablespoon olive oil
1 tablespoon minced fresh garlic
1 tablespoon finely chopped dried
    mushrooms

2 teaspoons ground coriander
2 teaspoons minced gingerroot
1/4 teaspoon ground cloves
1 bay leaf
8 cups vegetable broth
1 teaspoon salt
1 (8 3/4-oz.) can whole-kernel corn,
    drained
3 tablespoons minced fresh parsley

Combine the beans, barley, onion, oil, garlic, mushrooms, coriander, ginger, cloves, bay leaf and broth in a large slow cooker. Cook on low heat overnight, or until beans are tender.

Remove bay leaf. Stir in salt, corn and parsley. Soup can be thinned, if desired, by adding hot vegetable broth. Serve hot.

Makes 8 servings.

Variations

Add 1 tablespoon chili powder, 2 teaspoons ground cumin and a minced fresh or canned jalapeño chile to the soup for a Tex-Mex slant. Stir in 1/4 cup minced fresh cilantro in place of the parsley. Serve over rice, if desired.

Quick & Easy

Everyday Entrees

The emphasis is on fresh foods, and on quick dishes made mostly "from scratch." The exceptions to this are such time-savers as canned beans and prepared pasta sauces and salsas made with natural ingredients. Different cooking methods are used to prepare foods more quickly.

Stir-frying is a quick, nutritious and flavorful way to cook vegetarian main dishes. Because the vegetables are cooked quickly, they retain nutrients and color. Although the actual cooking takes just a few minutes, preparation of the vegetables can be time-consuming. And since the cook must be busy stirring while the food cooks, all of the vegetables must be ready beforehand.

Enter the microwave! Since cooking is stir-free, you can have some vegetables cooking as you prepare the rest. That means cooking and preparation take place simultaneously. That means, too, that your meal is on the table sooner, you'll be in a better mood, your digestion therefore will be enhanced, and, well, life just generally will be more pleasant all the way around!

Of course, the microwave is useful for cooking other dishes quickly, too. Often part of the cooking can be done in the microwave to make preparation quicker for dishes cooked in the oven or on the stovetop.

Also, remember that the broiler is a useful tool for quick pizzas and other dishes that cook quickly.

Vegetable-Topped Potatoes with Dijon-Cheddar Sauce

Potatoes are a comfort food for many of us. Topped with additional vegetables in a hearty sauce, they make a meal.

4 large russet potatoes, scrubbed
1 large onion, chopped
1 teaspoon canola oil
2 carrots, diced
2 celery stalks, thinly sliced
1 cup chopped broccoli flowerets

2 tablespoons butter or soy margarine
2 tablespoons whole-wheat flour
1/2 teaspoon Dijon mustard
1 1/4 cups low-fat milk
1/2 cup (2 oz.) shredded Cheddar cheese

Wrap each potato in waxed paper. Microwave potatoes on HIGH 10 minutes, or until tender.

Meanwhile, place the onion in a large microwave-safe bowl. Add oil and toss until combined. Microwave, uncovered, on HIGH 2 minutes. Stir in carrots. Microwave, uncovered, on HIGH 2 minutes more. Stir in celery. Microwave, uncovered, on HIGH 2 minutes. Add broccoli and microwave, uncovered, on HIGH 2 to 3 minutes, or until vegetables are firm-tender. Cover and set aside.

Melt margarine or butter in a medium saucepan over medium heat. Stir in flour and cook, stirring, 1 minute. Stir in mustard.

Add milk and bring to a simmer, stirring frequently. Cook, stirring, until sauce thickens. Add cheese and remove from heat. Stir until cheese is melted.

To serve, split baked potatoes and press open. Stir cheese sauce into vegetables. Spoon vegetable sauce over potatoes.

Makes 4 servings.

One-Ingredient-at-a-Time Microwave Oriental Stir-Fry

Arrange vegetables on your counter before you start. If the microwave time is completed before you have the next vegetable prepared, don't worry. The dish will stay warm, and you just continue to cook when you add the next vegetable. Preparing the rice and sauce first means everything comes together as soon as the veggies are done.

The general rule of thumb for microwave stir-fries is that longer-cooking vegetables go into the bowl first, then vegetables that require less cooking time are added. (I begin with onions, for while I like crisp snow peas or broccoli, I prefer my onions more tender.) You can mix and match whatever vegetables you have on hand.

1/4 cup vegetable broth	2 teaspoons canola oil
3 tablespoons pineapple juice	1 carrot
1 tablespoon soy sauce	1 medium stalk broccoli
1 1/2 teaspoons cornstarch	1/2 small yellow summer squash or
1 teaspoon honey	zucchini
1 large garlic clove, minced	1/2 yellow or red bell pepper
1/8 teaspoon ground ginger	1/2 cup frozen green peas
1 large onion, halved and thinly	1/2 cup chopped cashews
sliced crosswise	

Combine broth, juice, soy sauce, cornstarch, honey, garlic and ginger in a medium microwave-safe bowl and stir until combined. Microwave, uncovered, on HIGH 1 to 1 1/2 minutes, or until sauce is boiling and thickened. Stir and cover to keep warm.

Place the onion in a large microwave-safe bowl and toss with the oil. Microwave, uncovered, on HIGH 2 minutes.

Meanwhile, peel and thinly slice carrot on the diagonal. Stir carrot into onion and microwave, uncovered, on HIGH 1 minute.

Peel broccoli stalk. Thinly slice stem on the diagonal. Chop top. Stir into onion mixture and microwave, uncovered, on HIGH 1 minute.

Thinly slice squash on the diagonal. Stir into vegetables and microwave, uncovered, on HIGH 30 seconds.

Thinly slice bell pepper. Stir into vegetables and microwave, uncovered, on HIGH 30 seconds.

Stir in peas and sauce. Microwave, uncovered, on HIGH 30 seconds. Stir in cashews and serve over rice.

Makes 2 or 3 servings.

Anne's Hurry-Up Microwave Dinner

While the macaroni is cooking, prepare the remaining ingredients and assemble the casserole. A brief microwave heating, and you're ready to eat.

2 1/2 cups macaroni
1 onion, halved and thinly sliced
    crosswise
1/2 teaspoon olive oil
1 carrot, thinly sliced
1/4 cup finely diced red bell pepper

1 2/3 cups cooked (or 15-oz. can)
    black beans, drained
1 cup tomato sauce
1 cup low-fat milk
2 tablespoons minced fresh parsley
1 cup (4 oz.) shredded sharp
    Cheddar cheese

Cook the macaroni in a large pot of boiling water until firm-tender, about 5 to 6 minutes. Drain.

Meanwhile, combine the onion and oil in a medium bowl. Microwave, uncovered, on HIGH 2 minutes. Stir in carrot. Microwave, uncovered, on HIGH 1 minute. Stir in pepper. Microwave, uncovered, on HIGH for 2 minutes.

Place the drained macaroni, the vegetable mixture, beans, tomato sauce, milk and parsley in a 13 x 9-inch shallow, microwave-safe casserole dish. Microwave, uncovered, on HIGH 3 to 4 minutes, or until heated through. Stir macaroni and vegetables. Sprinkle with cheese. Microwave, uncovered, 2 minutes, or until cheese is melted.

Makes 6 servings.

Enchiladas for Two

Blend your own quick sauce with fresh or canned tomatoes. Stack tortillas with onions and cheese, heat and enjoy!

3 cups peeled, seeded and diced
    tomatoes
2 tablespoons tomato paste
2 garlic cloves, minced
1/2 jalapeño chile, seeded and minced
1 teaspoon chili powder
1/2 teaspoon ground coriander

1/2 teaspoon ground cumin
1/2 teaspoon salt
4 corn tortillas
1 cup finely chopped onion
3/4 cup (3 oz.) shredded sharp
    Cheddar cheese

Combine the tomatoes, tomato paste, garlic, chile, chili powder, coriander, cumin and salt in a blender. Process until smooth. Pour sauce into a medium saucepan over medium heat and bring to a boil. Reduce heat, cover and simmer 15 minutes.

To assemble casserole, lightly oil a shallow 9-inch-square microwave-safe casserole dish. Cover the bottom of the dish with a thin layer of sauce. Place a tortilla in the casserole dish and top with 1/3 of the onion and 1/4 of the cheese. Spoon on some of the sauce.

Repeat with 2 more tortillas, topping with onion, cheese and sauce. Add remaining tortilla to top, pour over remaining sauce and top with remaining cheese. Microwave, uncovered, on HIGH 4 minutes, or until heated through. The onion should stay almost crisp-tender.

Makes 2 servings.

Pronto Burritos

When you want a super-quick meal, look no further! Prepare the toppings while the filling heats up.

1 2/3 cups cooked (or 15-oz. can) pink beans, drained
1 2/3 cups cooked (or 15-oz. can) pinto beans, drained
2 (14 1/2-oz.) cans Mexican-style stewed tomatoes
1 1/2 cups frozen whole-kernel corn
1/4 head iceberg lettuce
1 cup cherry tomatoes
1/4 cup pitted ripe olives
2 green onions
1/4 cup fresh parsley
8 (7-inch) flour tortillas
1 cup (4 oz.) shredded Cheddar cheese
3/4 cup fat-free sour cream
Salsa

Place beans in a large saucepan and mash slightly with a potato masher. Place stewed tomatoes in a blender and process until smooth. Add half of the tomatoes to the beans and stir in corn. Cook bean mixture over medium-low heat, stirring occasionally, until heated through, about 5 to 6 minutes.

Meanwhile, thinly slice lettuce and place in a serving bowl. Halve cherry tomatoes, chop olives, thinly slice green onions and mince parsley. Place each in a separate serving bowl. Arrange tortillas, overlapping slightly, on a microwave-safe plate and warm in the microwave 1 minute on HIGH.

To assemble burritos, remove bean mixture from heat and stir in 3/4 cup of the shredded cheese. Fill each of the burritos with 1/8 of the bean mixture and roll up. Place seam-side down in a 13 x 9-inch microwave-safe casserole dish. Top with remaining blended tomatoes and sprinkle with remaining cheese. Microwave, uncovered, on HIGH 4 minutes, or until cheese is melted.

Spoon sour cream into a serving bowl, and arrange bowls of salsa, lettuce, cherry tomatoes, olives, green onions and parsley on the table. Let each person top the burritos as desired.

Makes 4 servings.

Madras Potatoes, Garbanzos & Peas

Garbanzos, or chickpeas, add extra protein to this hearty stovetop dish. Red kidney beans can be substituted for chickpeas. Serve with Indian Chapati Flat Bread (see page 199), if desired, and salad.

4 cups cubed, peeled potatoes
2 tablespoons canola oil
1 cup chopped onion
2 garlic cloves, minced
2 tablespoons whole-wheat flour
2 teaspoons Madras curry powder
1 cup vegetable broth
1 cup peeled, seeded and chopped
   tomatoes

1/2 cup cooked chickpeas
   (garbanzos)
1/2 cup frozen green peas
1/2 teaspoon salt
1/8 teaspoon freshly ground black
   pepper

Bring the potatoes to a boil in enough water to cover in a large saucepan over medium-high heat. Reduce heat and cook, uncovered, 8 minutes, or until the potatoes are firm-tender. Drain.

Meanwhile, heat the oil in a nonstick Dutch oven over medium heat. Add the onion and cook, stirring occasionally, until the onion is translucent, about 2 to 3 minutes. Add garlic, flour and curry powder and cook, stirring, 1 minute.

Add vegetable broth and stir until well combined with onion mixture. Bring to a boil and cook, stirring, until the mixture is thickened. Add cooked potatoes, tomatoes, chickpeas, peas, salt and pepper and return to a boil. Reduce heat, cover and simmer 5 minutes, or until flavors are blended.

Makes 4 servings.

Education

Accompaniments can add important protein and nutrients to a main dish. In the combination above, for example, the beans and peas in the main dish will complement the protein in the whole-grain flat bread. Salad adds a contrasting flavor and texture to the meal.

Jicama-Potato Skillet Cakes with Sour Cream Salsa

Jicama provides the pleasant texture of these attractive patties which are accompanied by a creamy salsa topping. Corn on the cob and garden green beans are great side dishes.

2 to 3 tablespoons grated onion	1/2 teaspoon salt
2 cups shredded jicama (see Note, below)	Dash of freshly ground black pepper
1 cup shredded russet potato	About 1 tablespoon canola oil
2 eggs, beaten	1 1/2 cups salsa
3 tablespoons unbleached all-purpose flour	1/4 cup fat-free sour cream
	1 teaspoon minced fresh cilantro

Squeeze liquid from the onion with your hand and place onion in a large bowl. Wrap the jicama in paper towels or cheesecloth and squeeze out the liquid. Do the same with the potato. (Russet potatoes are more starchy than red potatoes and are preferable for this dish.) Place in the bowl with the onion. Stir in the beaten eggs and sprinkle with flour, salt and pepper. Stir again until combined. Form jicama mixture into 8 cakes.

Heat oil in a large nonstick skillet over medium heat. Add patties, half at a time, and cook until browned, turning once, about 10 minutes. Keep patties hot in a warmed oven until all are cooked.

Combine salsa, sour cream and cilantro in a small bowl. Place some on skillet cakes and serve remainder on the side.

Makes 4 servings.

Note Jicama is a large root vegetable with a mild flavor and crisp texture that is eaten in Mexico and the Southwest. It can be eaten cooked or raw. Buy jicama that are smooth and firm. Do not refrigerate.

Stuffed Mushroom Caps
with Spinach Marrakech

Spinach and couscous get a Middle Eastern flavor with aromatic spices. Microwaving gets things done fast . . .

1/4 cup couscous
1 tablespoon fresh lemon juice
1 tablespoon water
2 teaspoons olive oil
1 teaspoon soy sauce
12 large mushrooms
1 cup packed spinach leaves, stems
   removed

2 tablespoons minced fresh parsley
1 teaspoon minced fresh mint
   (optional)
1 garlic clove, minced
1/4 teaspoon ground coriander
1/8 teaspoon paprika
Dash of ground fennel

Place the couscous, lemon juice, water, olive oil and soy sauce in a small bowl and stir until combined. Set aside.

Brush dirt from mushrooms and remove stems. Trim and discard tough ends, and finely chop stems. Add chopped stems to couscous mixture. Place the mushroom caps stem side down in a shallow, 13 x 9-inch microwave-safe baking dish and cover with waxed paper. Microwave on HIGH 4 minutes, rotating the dish once. Turn caps over.

Meanwhile, carefully rinse spinach. Shake off excess water and place in a medium nonstick skillet with the parsley and mint, if using, and garlic. Cook over medium-low heat, stirring frequently, until spinach is wilted. Drain in a sieve, pressing out any water with the back of a spoon. Finely chop spinach on a cutting board.

Combine the spinach, couscous mixture, coriander, paprika and fennel. Divide the mixture among the mushroom caps, filling them generously. Cover with waxed paper. Microwave on HIGH 6 minutes, turning once, or until mushrooms are tender.

Makes 4 servings.

Education

Low in calories, mushrooms are surprisingly nutritious. Rich in protein, with a good supply of B vitamins, mushrooms are also rich in minerals. When buying mushrooms, avoid those that appear shiny, bruised or pitted.

Fresh Herb & Tomato Pita Pizzas

Simple summertime fare. In winter, use 1 1/2 cups diced canned tomatoes in place of fresh and a sprinkling of dried marjoram and basil if no fresh herbs are available.

4 whole-wheat pita breads
2 cups (8 oz.) shredded mozzarella
    cheese
2 large tomatoes, peeled, seeded,
    finely chopped and drained

1 1/2 tablespoons minced fresh herbs
    (basil, marjoram, parsley,
    thyme or oregano in
    combination)

Preheat broiler. Place the pita breads on a baking sheet. Broil the pita bread about 1 to 2 minutes, until one side is warmed and slightly crisp. Turn and sprinkle about 1/3 cup of the mozzarella on each pita. Place under the broiler until cheese is melted.

Top with tomatoes. Place under broiler until heated through. Sprinkle with herbs and remaining cheese. Place under broiler until cheese is melted.

Makes 4 servings.

Education

Use fresh tomatoes to cut down on sodium. Most canned tomato products have sodium added.

West Coast Bean Burritos

The West Coast always seems to have more of everything—more rain (in the north), more fog (in the middle), and more sun (in the south). These burritos have more, too. Not just refried beans, but vegetables pack these hearty, tortilla-wrapped bundles. Serve with a green salad.

1 cup finely chopped onion
1 tablespoon olive oil
3 garlic cloves, minced
1 jalapeño chile, seeded and minced
1 teaspoon ground cumin
1 teaspoon ground coriander
1/2 teaspoon salt
1 cup shredded zucchini
1 cup shredded carrots
1 (16-oz.) can vegetarian refried
    beans

8 (7-inch) flour tortillas
1 1/2 cups (6-oz.) shredded sharp
    Cheddar cheese
1 (14 1/2-oz.) can Mexican-style
    stewed tomatoes
1 1/2 cups peeled, seeded and finely
    diced tomatoes
1/2 cup fat-free sour cream
1/4 cup nonfat yogurt
1 tablespoon minced fresh cilantro

Preheat oven to 400F (205C).

Combine the onion and oil in a large microwave-safe bowl. Microwave, uncovered, on HIGH 2 minutes. Stir in garlic, chile, cumin, coriander and salt. Microwave, uncovered, on HIGH 1 minute.

Stir in zucchini and carrots. Microwave, uncovered, on HIGH 2 minutes. Stir well. Microwave, uncovered, on HIGH 2 minutes, until vegetables are firm-tender. Stir in refried beans.

Warm tortillas on a microwave-safe plate on HIGH 30 seconds. To fill burritos, spoon 1/3 to 1/2 cup vegetable-bean mixture down the center of each tortilla. Sprinkle with some shredded cheese. Roll up and place filled tortillas seam side down in a lightly oiled, nonstick 13 x 9-inch baking pan.

Pour the stewed tomatoes into a blender and process until pureed. Spoon over the burritos. Top with fresh tomatoes. Bake in preheated oven 10 minutes, or until heated through.

Combine the sour cream, yogurt and cilantro in a small bowl. Add a dollop on the side of each serving.

Makes 4 servings.

Education

Shredding vegetables can save you time two ways. One, it is often faster than dicing, chopping or cutting vegetables into thin strips by hand. Two, shredded vegetables cook more quickly. To speed up dinner delivery, get out your shredder!

Spaghetti Squash with Garden Vegetables

A delightful juxtaposition of colors and flavors. A simple green salad and crusty bread would be excellent accompaniments.

1 small spaghetti squash
1 teaspoon olive oil
4 green onions, thinly sliced on
    diagonal
1 medium zucchini, diced
2 garlic cloves, minced
3 large, garden-ripe tomatoes,
    peeled, seeded and chopped
1/4 cup minced fresh parsley

2 tablespoons thinly sliced basil
    leaves
1 tablespoon drained capers (optional)
1 teaspoon balsamic vinegar
1/2 teaspoon salt
1/8 teaspoon freshly ground black
    pepper
1/3 cup freshly grated Parmesan
    cheese

Cut the spaghetti squash lengthwise in half. Place each half, cut side down, on a plate. Microwave on HIGH 5 to 6 minutes, or until tender. Keep warm.

Meanwhile, heat the oil in a medium, nonstick skillet over medium heat. Add the green onions and cook, stirring frequently, until the onions are translucent, about 1 to 2 minutes. Add the zucchini and garlic and cook, stirring, 1 to 2 minutes, or until the zucchini is just crisp-tender.

Add tomatoes, parsley, basil, capers, if using, vinegar, salt and pepper and bring to a boil. Reduce heat, cover and simmer 5 minutes.

To serve, remove seeds from squash halves. With a fork, gently tease out strands of squash. Serve on warmed plates topped with garden vegetable mixture. Sprinkle with cheese.

Makes 4 servings.

Mexican Pita Pizza

Salsa adds a piquant highlight to this "fast-lane" main dish. Make two servings in well under 10 minutes! These are also a wonderful, and wholesome, snack for kids.

2 whole-wheat pita breads
2/3 cup (about 2 1/2 oz.) shredded
    sharp Cheddar cheese
1 cup Garlic-Lovers' Salsa Cruda
    (see page 3) or bottled salsa

1/4 cup pitted ripe olives
Fresh cilantro leaves, for garnish

Preheat broiler. Place the pita breads on a baking sheet. Broil the pita bread about 1 to 2 minutes, until one side is warmed and slightly crisp. Turn over and sprinkle cheese on remaining side. Broil until cheese is melted.

Meanwhile, drain salsa of excess liquid. Chop ripe olives and toss with salsa. Divide salsa mixture and spread over melted cheese. Heat under broiler and serve, garnished with cilantro.

Makes 2 servings.

Education

Having a quick and healthy main dish in your recipe repertoire makes it more likely you'll eat well. When you can prepare a light dinner in minutes, it is less likely you'll turn to junk food.

Pasta with Goat Cheese, Peppers & Pine Nuts

Goat cheese comes in a variety of textures, from rich and creamy to dry and crumbly. The drier cheese works best here. Blue cheese can also be used.

3/4 pound fusilli or linguine
2 tablespoons olive oil
1 onion, halved and thinly sliced
    crosswise
1 cup diced summer squash
1 cup diced zucchini
2 garlic cloves, minced

1 (7-oz.) jar roasted red bell peppers,
    drained and cut into thin,
    2-inch strips
1 cup (about 4 oz.) crumbled goat
    cheese
1/4 cup pine nuts
1 tablespoon minced fresh parsley

Bring a large pot of water to a boil. Add pasta and cook 7 to 10 minutes, or until firm-tender. Drain.

Meanwhile, heat the oil in a large, nonstick skillet over medium heat. Add the onion, summer squash and zucchini and cook, stirring frequently, until vegetables are crisp-tender, about 5 to 6 minutes. Add garlic and stir to combine. Add bell peppers and cook, stirring frequently, 2 to 3 minutes or until heated through.

To serve, toss hot cooked pasta, vegetables, cheese and pine nuts together in a warmed serving bowl. Sprinkle with parsley.

Makes 4 servings.

Tijuana Tacos

Anything goes with a round-the-world taco filling that includes bulgur (for a marvelous, chewy texture), carrots, onions and your choice of beans. The filling is versatile, too. See variations, below.

18 taco shells
1 tablespoon olive oil
2 onions, chopped
2 carrots, finely diced
2 garlic cloves, minced
1/2 cup bulgur
1 tablespoon chili powder
2 teaspoons ground coriander
2 teaspoons ground cumin
2 1/2 cups eight-vegetable juice

1 2/3 cups cooked (or 15-oz. can)
    Roman or pinto beans, drained
1/4 cup minced fresh cilantro
1 1/2 cups (6 oz.) shredded sharp
    Cheddar cheese
1 cup shredded lettuce
1 cup diced tomatoes
3/4 cup finely chopped green onions
1/3 cup low-fat sour cream (optional)
1/3 cup non-fat yogurt (optional)

Preheat oven to 175F (85C). Unwrap taco shells and put in oven to warm.

Heat the oil in a large, nonstick skillet over medium heat. Add the onions and carrots and cook, stirring occasionally, until the vegetables begin to brown, about 6 to 7 minutes. Stir in bulgur, chili powder, coriander and cumin and cook, stirring, 1 to 2 minutes.

Stir in juice and bring to a boil. Reduce heat, cover and simmer 15 minutes, or until the bulgur is tender. Add beans to pan and mash them slightly with the back of a wooden spoon. Stir in cilantro and heat through.

To serve tacos, add some of the bean mixture to a taco. Toppings can be selected from cheese, lettuce, tomatoes and green onions. Stir together the sour cream and yogurt. Guests can drizzle this over top, if desired.

Makes 8 to 10 servings.

Variations

Tijuana Burritos Fill flour tortillas with bean mixture, roll and place seam side down in a baking dish. Brush liberally with salsa, sprinkle with a little shredded Cheddar cheese, and cover with foil. Bake at 350F (175C) 15 minutes, or until heated through and cheese is melted.

Tijuana Burgers Add 2 to 3 teaspoons of whole-wheat flour for each 1/2 cup of mixture. Form into patties and cook in a lightly oiled, nonstick skillet. A great way to use up leftover filling. Delicious with Farmhouse Home Fries (page 192).

Pronto Dinner Cheese Nachos

On those casual Sunday nights on the deck or in front of the TV, or when no one is particularly hungry for a big meal, nachos come to the rescue. Accompanied by a salad, they make a light dinner without a lot of work.

Low-fat tortilla chips
1 cup vegetarian refried beans
1/2 cup tomato juice
1/4 cup salsa
1 garlic clove, minced

1/4 teaspoon ground coriander
1/4 teaspoon ground cumin
1 1/2 cups (6 oz.) shredded sharp
    Cheddar cheese

Preheat broiler. Arrange a layer of tortilla chips, slightly overlapping, on a large baking sheet with edges. (That way cheese won't melt onto the oven floor.)

In a small saucepan, combine the refried beans (or any mashed beans), tomato juice, salsa (dice any large chunks of ingredients), garlic, coriander and cumin. Bring to a boil over medium heat.

Spoon the beans over the tortilla chips. Sprinkle with cheese. Broil just until cheese is melted.

Makes 4 servings.

Education

Complete proteins provide all of the essential amino acids needed by our bodies. Legumes, like the beans, above, are rich in lysine, but poor in methionine. Add corn chips, however, and the protein is balanced. When eaten together, corn and beans "complete" the protein picture.

Stir-Fried Cabbage with Snow Peas

A quick dinner for two. Prepare or reheat rice, then quickly stir-fry the remaining ingredients.

1 tablespoon olive oil
2 tablespoons minced onion
4 cups chopped cabbage
1 cup snow peas, stems and strings
    removed, halved diagonally
1 teaspoon shredded gingerroot

1/2 teaspoon minced garlic
1/2 cup vegetable broth
1 tablespoon soy sauce
1 1/2 teaspoons cornstarch
1 tablespoon chopped cashews

Heat the oil in a large nonstick skillet over medium heat. Add the onion and cabbage and cook, stirring frequently, until cabbage is crisp-tender, about 3 to 4 minutes. Add snow peas, gingerroot and garlic and cook, stirring, 2 to 3 minutes, until peas are bright green and crisp-tender.

Combine broth, soy sauce and cornstarch in a measuring cup. Add to cabbage mixture and cook, stirring, until sauce is thickened, about 1 to 2 minutes. Stir in cashews and serve over rice.

Makes 2 servings.

Education

Surprisingly, fresh peas are an excellent source of protein. Like other legumes, they are also low in fat and calories—but unlike dried beans, they cook quickly. Snow peas have a little less protein than mature, shelled peas, but are higher in calcium, iron and vitamin C.

Minted Vegetable Stir-Fry with Cashews

Served over quick brown rice, minted vegetables create a light, but satisfying summer main dish. Cashews complement rice for added protein.

2 1/4 cups water
2 cups instant brown rice
1/4 teaspoon salt
1 tablespoon canola oil
1 medium red onion, halved and thinly sliced crosswise
4 medium carrots, cut on the diagonal into 1/4-inch-thick slices
1 red bell pepper, cut into thin strips

3 cups thinly sliced red cabbage
3 cups coarsely chopped broccoli flowerets
1/2 cup vegetable broth
2 tablespoons soy sauce
3 garlic cloves, thinly sliced
1/3 cup chopped cashews
3 tablespoons minced fresh mint leaves

Bring the water to a boil in a medium saucepan. Stir in rice and salt. Reduce heat, cover and simmer 10 minutes, or until water is absorbed. Remove from heat and keep covered until ready to use.

Heat the oil in a nonstick wok or Dutch oven over medium heat. Add the onion, carrots and bell pepper and cook, stirring frequently, until the onion is crisp-tender, about 5 to 6 minutes. Add cabbage, broccoli, broth, soy sauce and garlic and cook, stirring, until broccoli is crisp-tender, 2 to 3 minutes. Stir in cashews and mint.

Fluff rice and serve topped with vegetables.

Makes 4 servings.

Education

Although similar in flavor to green cabbage, red cabbage has more vitamin C. So much, in fact, that one 3 1/2-ounce serving provides nearly 100 percent of the Recommended Daily Allowance.

Kashmiri Cauliflower

Serve a pleasantly curried vegetable combination over brown rice. Indian Chapati Flat Bread, an easy-to-make stovetop flat bread (see page 199), is a great accompaniment.

2 1/4 cups vegetable broth
2 cups instant brown rice
1 teaspoon butter or soy margarine
2 tablespoons canola oil
2 onions, halved and thinly sliced
2 teaspoons ground coriander
2 teaspoons ground fennel seeds
1 teaspoon ground cumin

1 teaspoon whole mustard seeds
1/2 teaspoon salt
Dash of ground cayenne
4 cups finely chopped cauliflowerets
1/3 cup water
2 cups diced canned tomatoes with juice
2 tablespoons fresh lemon juice

Bring the broth to a boil in a medium saucepan. Stir in rice and butter. Reduce heat, cover and simmer 10 minutes, or until water is absorbed. Remove from heat and keep covered until ready to use.

Heat the oil in a nonstick Dutch oven over medium-high heat. Add the onions and cook, stirring frequently, until the onions are tender, about 5 to 6 minutes. Add coriander, fennel, cumin, mustard seeds, salt and cayenne and cook, stirring, 1 minute.

Add cauliflower and stir until well combined with spices. Add water and stir. When mixture comes to a boil, reduce heat, cover, and simmer, stirring occasionally, until the cauliflower is firm-tender, about 6 minutes. (Add a few extra drops of water if necessary; do not let the cauliflower become mushy.)

Add the tomatoes and lemon juice and simmer until the flavors are combined, about 5 minutes. Serve over rice.

Makes 4 servings.

Education

Cook cauliflower too long, and in addition to a loss in flavor and texture, you will deplete its stores of folacin. In fact, tests seem to show that folacin in cauliflower is more quickly lost to the water in which it is cooked than is the case for any other vegetable. When the cauliflower is cooked in water, you can conserve the folacin by including this liquid in the finished dish, as above.

Zucchini Pancakes with Herbed Tomatoes

A summer garden highlight, if ever there was one! This dish was a real hit with friends.

1 pound zucchini, shredded
1/2 cup (2 oz.) shredded fontina or
    mild Cheddar cheese
1/2 cup whole-wheat flour
1 tablespoon minced fresh parsley
2 eggs, beaten
1/4 teaspoon salt

About 4 teaspoons olive oil
1/2 cup thinly sliced green onions
2 cups peeled, seeded and diced
    tomatoes
1 tablespoon thinly sliced basil
    leaves or 1 1/2 teaspoons dried
    basil

Combine the zucchini, cheese, flour and parsley in a large bowl. Pour in eggs and salt and stir until well combined.

Heat 2 teaspoons oil in a griddle or large skillet over medium heat. Spoon batter into small pancakes. When bottoms are set and browned, turn to brown remaining side. Cook until browned and cooked through, about 10 minutes total.

Meanwhile, heat remaining oil in a medium saucepan over medium heat. Add the green onions and cook, stirring, just until translucent, about 1 minute. Add tomatoes and just heat through, stirring occasionally. Remove from heat and stir in basil.

Serve zucchini pancakes on a warmed serving platter with tomatoes over top or on the side.

Makes 4 servings.

Education

Zucchini is a perennial partner with tomatoes. A tomato topping here gives the authority of a main dish to the zucchini pancakes. Eggs provide the protein, without overdoing fat.

Middle Eastern Eggplant Steaks

Thick slices of broiled eggplant are topped with flavorful feta cheese sauce. Choose an eggplant that is long and lean rather than rounded. Serve over couscous, rice or orzo pasta.

1 large eggplant
2 garlic cloves, halved
1 tablespoon plus 2 teaspoons fresh
    lemon juice
1 tablespoon olive oil
1 cup low-fat yogurt
1/2 cup crumbled feta cheese

1/4 cup finely chopped red onion
3 tablespoons minced fresh mint or
    cilantro
1 tablespoon minced fresh parsley
2 garlic cloves, minced
Mint sprigs, for garnish
Tomato wedges, for garnish

Preheat broiler. Trim off eggplant stem and blossom ends and peel. Cut lengthwise into slices, about 3/4 inch thick. Rub the slices with the garlic halves. Discard garlic. Combine 2 teaspoons of the lemon juice and the olive oil. Brush mixture on both sides of eggplant slices.

Place slices in one layer on a baking sheet. Broil 6 to 8 minutes, or until tender, turning once to brown both sides.

Meanwhile, combine yogurt, feta, onion, minced mint or cilantro, parsley, remaining tablespoon lemon juice and garlic in a small saucepan. Set over very low heat and warm gently. Do not boil.

To serve, top eggplant steaks with yogurt sauce. Garnish with fresh mint sprigs and tomato wedges.

Makes 4 servings.

Education

Feta cheese adds lots of flavor to dishes, but is high in salt. By combining feta with lower-sodium ingredients, such as yogurt, you can enjoy feta's distinctive flavor without getting too much sodium in your diet.

Salsa Surprise

Have some rice, corn and canned beans on hand? Use salsa to flavor a quick casserole. Run the cheese topping under the broiler, and you're ready to serve!

2 1/4 cups vegetable broth or water
2 cups instant brown rice
1/4 cup minced onion
1 tablespoon butter or soy margarine
1/2 teaspoon salt
1/4 teaspoon ground coriander
1 2/3 cups cooked (or 15-oz. can)
    kidney, pinto or pink beans,
    drained

1/2 cup canned or frozen whole-
    kernel corn
1 cup salsa
3/4 cup (3 oz.) shredded Monterey
    Jack cheese

Bring the stock to a boil in a large saucepan over medium-high heat. Stir in rice, onion, butter, salt and coriander and return to a boil. Reduce heat, cover and simmer 10 minutes, or until water is absorbed.

Stir the beans, corn and salsa into the rice. Cover and steam an additional 2 to 3 minutes, or until heated through. Turn rice mixture into a 13 x 9-inch shallow casserole dish, spreading evenly without packing. Sprinkle with cheese and place under broiler until the cheese is melted.

Makes 4 servings.

Education

Salsa is a quick and easy flavoring that can be added to a variety of vegetarian dishes. Two additional advantages, besides its convenience, is that salsa adds no fat to a dish and the vitamin C of the tomatoes and chiles or bell peppers facilitates iron absorption from any beans in the dish.

Angel Hair Pasta with Vegetable Cheese Sauce

Angel hair pasta cooks quickly and does not hold well. For that reason, the sauce should be nearly completed before adding the pasta to boiling water.

1 tablespoon olive oil
1 tablespoon butter or soy
    margarine
1 large onion, chopped
2 medium carrots, diced
1 small zucchini, diced
1 cup quartered mushrooms
2 tablespoons whole-wheat flour
1 1/2 cups low-fat milk

1/4 cup tomato sauce
1/2 pound angel hair pasta, broken
    in half
1/3 cup (about 1 1/2 oz.) shredded
    sharp Cheddar cheese
1 tablespoon freshly grated
    Parmesan cheese
1 tablespoon minced fresh parsley

Bring a large pot of water to a boil, but do not add pasta until sauce is nearly completed.

Meanwhile, heat the oil and butter in a nonstick Dutch oven over medium heat. Add the onion and carrots and cook, stirring frequently, until the onion is translucent, about 3 to 4 minutes. Add zucchini and mushrooms and cook, stirring frequently, until the vegetables are firm-tender, about 4 to 5 minutes.

Stir in flour until combined with vegetables. Add milk and tomato sauce and cook, stirring, until mixture begins to thicken and comes to a boil. Reduce heat.

Add pasta to boiling water, and stir to separate until water returns to a boil. Reduce heat slightly and boil 2 to 3 minutes, or just until the pasta is firm-tender.

Meanwhile, simmer vegetable mixture 1 minute, stirring occasionally. Stir in cheeses and parsley. Remove pan from heat, cover and set aside.

Drain pasta. Divide pasta among warmed individual plates. Stir vegetable mixture and spoon over servings of pasta.

Makes 4 servings.

Acorn Squash Stuffed with Golden Couscous

The microwave makes short work of preparing winter squash, the basis of this delightful autumnal main dish.

2 large acorn squash
1 small onion
1 tablespoon butter or soy margarine
1/3 cup slivered almonds or chopped
  cashews
1/3 cup golden or regular raisins
1/2 teaspoon ground coriander
1/4 teaspoon ground cinnamon

1/8 teaspoon ground turmeric
Dash of ground cardamom
2/3 cup couscous
1 cup boiling water
1/4 cup low-fat yogurt
1/4 cup fat-free sour cream
1/2 teaspoon maple syrup
Mint sprigs, for garnish

Halve squash lengthwise and remove seeds. Place cut side up in a microwave-safe 13 x 9-inch casserole dish containing 1/2 inch of water in the bottom. Microwave, uncovered, on HIGH 12 to 13 minutes, or until squash are tender, rotating the squash halves once to ensure even cooking.

Meanwhile, cut the onion in half lengthwise and thinly slice crosswise. Melt the butter in a medium nonstick skillet over medium heat. Add the onion and cook, stirring occasionally, until the onion is nearly tender, about 3 to 4 minutes. Stir in nuts and cook, stirring, 1 minute until nuts are lightly toasted. Add raisins, coriander, cinnamon, turmeric and cardamom and cook, stirring, 1 minute.

Stir in couscous until combined with onion mixture. Add boiling water, cover, and remove from heat. Let stand 5 minutes.

To assemble, divide the couscous filling between the squash halves. Cover with waxed paper and microwave on HIGH 3 minutes, until heated through. Combine yogurt, sour cream and maple syrup in a small microwave-safe dish. Microwave on HIGH 30 to 45 seconds, or just until heated through. Stir well and drizzle 2 tablespoons of the sauce over each squash half. Garnish with mint sprigs.

Makes 4 servings.

Fettuccine "Don't Be Afraid-O"

Loosely based on the famous dish fettuccine Alfredo (in that they both feature fettuccine . . .), this version is low in fat and cholesterol, but outstandingly creamy and tasty! Fresh basil does the trick here!

1 pound fettuccine
1 cup low-fat ricotta cheese
1/2 cup low-fat cottage cheese
1/4 cup crumbled blue cheese
1 cup lightly packed basil leaves

1 large garlic clove, minced
Dash of salt
Dash of ground cayenne
Basil sprigs, for garnish

Bring a large pot of salted water to a boil. Add the fettuccine and cook until pasta is firm-tender, about 9 to 11 minutes. Drain pasta, reserving 1/3 cup of the pasta water.

Meanwhile, place the cheeses in a food processor and process until smooth. Add basil leaves, garlic, salt and cayenne. Process again, just until combined.

Add pasta water to the cheese mixture and process until smooth. Drain fettuccine and place in a warmed serving bowl. Toss with cheese mixture, and garnish with basil sprigs. Serve at once.

Makes 6 servings.

Education

Choose low-fat dairy products to take advantage of their strong points (a rich supply of protein and calcium) without getting too much fat. With the choices in dairy products available these days—reduced fat, low-fat and fat-free—you can custom-tailor your dishes to your own needs.

Marco Polo Spaghetti

An easy, tasty and colorful dish. Don't brown the garlic, or the taste will become bitter.

3/4 pound spaghetti
2 tablespoons olive oil
3 garlic cloves, minced
2/3 cup chopped walnuts
1/2 cup sliced ripe olives

1 (4-oz.) jar pimientos, drained and
    chopped
1/3 cup minced fresh parsley
1/4 cup thinly sliced fresh basil leaves
1/4 cup freshly grated Parmesan
    cheese

Bring a large pot of water to a boil. Add the spaghetti and cook 5 to 7 minutes, or until firm-tender. Drain.

Meanwhile, heat the oil in a small skillet over medium-low heat. Add the garlic and cook, stirring constantly, about 1 minute. Stir in walnuts and cook 1 minute, or just until the garlic begins to turn golden.

When the spaghetti is drained, return it to the empty pot in which it was cooked. Spoon garlic mixture over spaghetti and toss. Add olives, pimientos, parsley and basil and toss again. Top with freshly grated Parmesan cheese.

Makes 4 servings.

Pesto Genovese

An Italian sauce that takes the glories of fresh basil to their peak! It is made easily with a food processor instead of the traditional mortar and pestle. Toss some of this sauce with hot pasta or cooked vegetables and your taste buds will know bliss.

6 garlic cloves, peeled
1/2 teaspoon salt
1/2 cup pine nuts
1/3 cup freshly grated Parmesan cheese

1/4 cup extra-virgin olive oil
1/4 cup boiling water
2 cups fresh basil leaves

Place the garlic and salt in a food processor and process until garlic is finely chopped. Add pine nuts and process until combined. Add cheese and process until combined, scraping down the sides of the bowl as necessary.

Add boiling water and process until smooth. (This tempers the raw taste of the garlic and creates a more mellow sauce.) Add basil leaves and process until combined, scraping down the sides of the bowl.

Makes about 1 1/4 cups.

Variations

Toss pesto with cooked tortellini or gnocchi, with or without cooked green peas. Toss pesto with spaghetti. Substitute pesto for salad dressing for hot potato salad.

Spaghetti with Olives, Walnuts & Tomatoes

When summer yields the most succulent, ripe tomatoes is the time to serve this fresh tomato pasta. Mixing green and ripe olives enriches the flavor.

3/4 pound thin spaghetti
1 tablespoon olive oil
1 garlic clove, pushed through a
    press
2 large, ripe tomatoes, peeled,
    seeded and finely chopped
1/3 cup chopped walnuts

1/4 cup sliced pitted ripe olives
1/4 cup fresh basil leaves, thinly
    sliced
1 tablespoon sliced green olives
    with pimiento
Dash of freshly ground pepper

Bring a large pot of water to a boil. Add the spaghetti and cook 4 to 6 minutes, or until firm-tender.

Meanwhile, combine the oil and garlic in a small dish. Drain the spaghetti and place in a warmed serving bowl. Toss immediately with the oil and garlic.

Add the tomatoes, walnuts, ripe olives, basil, green olives and pepper and toss until combined.

Makes 4 servings.

Education

Nuts complement pasta (made from grain) for a high-protein main dish. Remember that nuts are high in fat, so a modest amount for each serving should be the rule.

Linguine with Basil & Grated Cheese

When using a rich-tasting cheese such as fresh Parmesan to enhance flavor, remember that a little goes a long way. A tomato and onion salad goes well with this dish.

1 pound linguine
2 tablespoons butter or soy margarine
1/3 cup freshly grated Parmesan
    cheese

16 large fresh basil leaves, thinly sliced

Bring a large pot of water to a boil. Add the linguine and cook 8 to 10 minutes, or until firm-tender. Drain and place in a warmed serving bowl. Toss immediately with the butter or margarine until melted. Toss with cheese and basil.

Makes 4 servings.

Education

Fresh herbs are great additions to cooking that can lessen a cook's reliance on less healthy fat and salt to flavor foods.

Of course, the easier it is to reach herbs, the more likely you will be to use them. A planter or pot of herbs outside the kitchen door makes it easy to get fresh cuttings. Winter or summer, herbs can also be grown on an indoor windowsill.

Thin Spaghetti with Broccoli Rabe

A quick sauté of garlic and greens with spaghetti is finished off in the skillet and a most modest dusting of cheese is added—dinner for two.

1/3 pound thin spaghetti	2 tablespoons thinly sliced garlic
1 bunch (about 3/4 pound) broccoli rabe	1/4 cup freshly grated Parmesan cheese
2 tablespoons olive oil	

Bring a large pot of water to a boil.

Meanwhile, rinse broccoli rabe and trim off stems. Coarsely chop. Heat the oil in a large skillet. Add the garlic and cook over medium heat, stirring constantly, until the garlic is just beginning to turn golden at the edges; do not brown garlic or it will become bitter. Add broccoli rabe and cook, stirring, until greens are well wilted, about 2 to 3 minutes.

Add spaghetti to boiling water and cook about 5 minutes, or until quite firm-tender. Drain pasta, reserving 1/4 cup of the pasta water. Add pasta water and pasta to the skillet. Cook, stirring, until ingredients are combined.

Add cheese and cook, stirring, until cheese begins to melt. Serve immediately.

Makes 2 servings.

Education

Greens are an excellent source of vitamin C and beta carotene, and contain iron, calcium, folacin and fiber. A rule of thumb is, the darker the greens, the more nutritious. If you find broccoli rabe too bitter, substitute escarole in the dish above. A few shredded leaves of radicchio can be added for color.

Herbed Garlic Rigati with Fresh Tomatoes

For a mild, delicious flavor, cook the garlic cloves with the pasta, then push through a press while finishing the dish. This dish makes a quick lunch for two, but can easily be doubled. It is best, of course, with garden-ripe tomatoes.

1 1/2 cups (about 1/4 pound) rigati
    or rigatoni pasta
2 garlic cloves, peeled
1 thin lemon slice
1 large tomato
1 tablespoon minced fresh parsley

1 tablespoon minced fresh basil
1/4 teaspoon dried marjoram,
    crumbled
1 teaspoon extra-virgin olive oil
Dash of salt
Dash of freshly ground pepper

Bring a large saucepan of water to a boil. Add pasta, garlic and lemon slice and cook 8 to 10 minutes, or until firm-tender.

Meanwhile, peel, seed and finely dice tomato. Combine in a bowl with the parsley, basil and marjoram.

Drain pasta and turn into a warmed serving bowl. Remove lemon slice and garlic cloves. Push garlic through a press into a small dish. Add garlic and oil to tomato mixture.

Spoon tomato mixture over hot pasta and toss until ingredients are combined. Season with salt and pepper. Serve hot or at room temperature.

Makes 2 servings.

Education

Long, moist cooking mellows the punch of garlic, making it more palatable to those who would rather avoid its pungent flavor. In the dish above, boiling the garlic with the pasta does the trick, adding garlic's healthy qualities to the dish without its bite.

Rosemary, Olive & Walnut Pasta

Fresh rosemary and a touch of fresh thyme complement the briny taste of green olives in a quick pasta sauce. The microwave makes short work of preparation right in the serving dish.

1 onion, chopped
1 teaspoon extra-virgin olive oil
2 garlic cloves, minced
2 cups canned chopped tomatoes
    with juice
1/4 cup walnuts, chopped

1/4 cup pitted green olives,
    chopped
1 teaspoon minced fresh rosemary
1/4 teaspoon fresh thyme leaves
3/4 pound angel hair pasta

Bring a large pot of water to a boil. Meanwhile, place the onion in a large, microwave-safe serving dish. Add oil and toss to combine. Microwave, uncovered, on HIGH 1 minute. Stir in garlic and microwave, uncovered, on HIGH 1 minute.

Add tomatoes, walnuts, olives, rosemary and thyme to onion. Toss to combine. Set aside.

When water is boiling, cook pasta just until firm-tender, about 2 to 3 minutes. Drain and toss with sauce.

Makes 4 servings.

Cheese Tortellini with Walnuts & Tomatoes

Buy fresh or frozen cheese tortellini, which may be available in spinach (green) or white pasta. Or buy some of each, and mix them each time you serve this dish.

1 pound cheese tortellini
1 tablespoon olive oil
3 garlic cloves, minced
2 tablespoons chopped walnuts
4 cups peeled, seeded and diced
    tomatoes or canned plum
    tomatoes
1/4 cup thinly sliced fresh basil leaves

2 tablespoons minced fresh parsley
1/4 cup freshly grated Parmesan
    cheese
1/4 teaspoon salt
Dash of cayenne
1/8 teaspoon freshly ground black
    pepper
Basil or parsley sprigs, for garnish

Cook tortellini in a large saucepan of boiling water according to package directions. Drain.

Meanwhile, heat the oil in a nonstick Dutch oven over medium heat. Add the garlic and walnuts and cook, stirring frequently, until the garlic just begins to turn golden; do not brown garlic or it will become bitter. Add tomatoes, basil and parsley and bring to a boil. Reduce heat, cover and simmer 5 minutes.

Add drained tortellini to the tomato mixture with the cheese, salt, cayenne and black pepper. Heat through and serve garnished with basil or parsley sprigs.

Makes 4 servings.

Hot from the Oven

Cooking dishes in the oven frees you to do other things while your food cooks. The selection here includes stuffed bell peppers and squash, casseroles and even pizza and an easy lasagne made with no-cook noodles. Dishes from the oven are particularly appealing when the weather is cooler and we feel the need for heartier food.

Some recipes call for ingredients to be cooked in the microwave or on the stovetop before being combined with the remaining ingredients, reducing the time in the oven.

Butternut Squash with Wild Rice

If you are looking for a feast without feathers during the holiday season, stuffing delicious butternut squash with festive cranberries, nuts and wild rice should fit the bill. The recipe leans ever so slightly toward the extravagant, without going overboard—perfect for a cele-bration!

4 small (about 1 1/4 lbs. each) butternut squash	1 cup chopped onion
1 cup wild rice	1 cup finely diced carrots
1/2 cup long-grain brown rice	1 cup chopped pecans
3 1/2 cups vegetable broth	1/2 cup finely chopped cranberries
1 tablespoon minced shallots	1/2 cup peeled, cored and finely diced apples
2 tablespoons butter or soy margarine	1/4 cup chopped raisins
	2 tablespoons minced fresh parsley

Preheat oven to 350F (175C). Cut the squash in half (see Note, below) and scoop out seeds and fibers. Place cut side up in shallow baking dishes, and tightly cover with foil. Bake 40 to 50 minutes, or until flesh is firm-tender.

Meanwhile, rinse the wild rice well. (This removes any smoky aroma.) Combine with brown rice, broth and shallots in a large saucepan and bring to a boil. Reduce heat, cover and simmer 45 minutes, or until nearly all the liquid is absorbed.

Melt butter in a nonstick Dutch oven over medium-high heat. Add the onion and carrot and cook, stirring occasionally, until the onion turns golden, about 7 to 8 minutes. Add pecans and cook, stirring occasionally, 2 to 3 minutes. Add cranberries, apples and raisins and cook, stirring occasionally, 3 to 4 minutes.

Remove butternut squash from oven and cool slightly. Scoop out pulp, leaving a firm, 1/2-inch shell. Dice the squash pulp. Add squash, cooked rice and parsley to the onion mixture in the Dutch oven, stirring until combined.

Divide the stuffing among the squash halves. Cover tightly with foil and return to oven 20 minutes, or until heated through.

Makes 8 servings.

Note To cut butternut squash, position the squash on a damp kitchen towel on a flat work surface. Fold up the sides of the towel to hold the squash in place. Center a large chef's knife over the center of the squash, lengthwise, and tap with a rubber mallet until squash is nearly halved. Finish cutting by hand.

Squash can also be softened slightly in the microwave before cutting. First, with a sharp knife, pierce the skin through to the cavity in several places. Microwave each squash 2 to 3 minutes, or until squash skin is slightly softened. Cut in half with a sharp knife.

Broccoli-Corn Cheese Custard

Arrange vegetable-rich, delicately flavored cheese custard over rice. Add a salad and sit down to dinner!

1 large onion, chopped
1 teaspoon olive oil
2 stalks broccoli, stems peeled, chopped (about 2 cups)
1 cup fresh or frozen whole-kernel corn
3 eggs
1 1/2 cups low-fat milk

3/4 cup (3 oz.) shredded Cheddar cheese
1/4 teaspoon salt
Dash of freshly ground black pepper
1/2 cup dry bread crumbs
1 tablespoon melted butter or soy margarine

Preheat oven to 350F (175C). Lightly oil a large soufflé dish or deep casserole dish.

Place the onion in a medium microwave-safe bowl. Add the oil and toss until combined. Microwave, uncovered, on HIGH 2 minutes. Stir in broccoli. Microwave, uncovered, on HIGH 2 minutes. Stir in corn. Microwave, uncovered, on HIGH 1 minute.

Beat the eggs in a large bowl and combine with the milk, cheese, salt and pepper. Stir in the broccoli mixture. Pour into prepared dish. Place the bread crumbs and melted butter in a small bowl and stir together until well combined. Sprinkle the crumbs along the outside edge of the casserole.

Place the casserole dish in a pan with 1 to 2 inches of water in it. Bake 50 to 60 minutes, or until the custard is set.

Makes 4 or 5 servings.

Oriental-Style Baked Potatoes

Baked potatoes provide a base for a variety of vegetarian main dishes. Topped with vegetables or chili, they are satisfying and full of flavor. The Oriental stir-fry topping here is also wonderful on rice, of course.

4 large russet potatoes
1/4 cup soy sauce
3 tablespoons honey
2 tablespoons rice vinegar
2 tablespoons tomato juice
2 teaspoons cornstarch
1 teaspoon finely minced garlic
1 teaspoon finely minced gingerroot
1/4 to 1/2 teaspoon crushed red
    pepper flakes, or to taste

1 tablespoon canola oil
1 large onion, halved and thinly
    sliced crosswise
1 large carrot, halved and thinly
    sliced crosswise
1 cup chopped broccoli flowerets
1 cup chopped cauliflowerets
1 cup snow peas, trimmed
1/3 cup thinly sliced celery

Preheat oven to 350F (175C). Scrub potatoes and pierce in several places with a sharp knife. (This prevents a buildup of steam, which could cause the potatoes to explode.) Bake until tender, about 45 minutes.

Meanwhile, combine soy sauce, honey, vinegar, tomato juice, cornstarch, garlic, ginger and pepper flakes in a small bowl. Stir well, and set aside.

Heat oil in a wok or nonstick Dutch oven over medium-high heat. Add the onion and carrot and cook, stirring constantly, until the onion is translucent, about 3 to 4 minutes.

Add broccoli, cauliflower, snow peas and celery and stir-fry until the ingredients are crisp-tender, about 5 to 6 minutes. Stir soy sauce mixture and add to wok. Cook, stirring, until the sauce has thickened, about 2 to 3 minutes.

To serve, split potatoes and spoon hot stir-fry mixture over each.

Makes 4 servings.

Education

The lowly potato is the most popular vegetable in the world. I'm reminded of a friend who, when hiking in Nepal, stayed at a tea house where the informal menu included potatoes and a vegetable of the day. Ordering both, my friend was served potatoes and more potatoes!

Just by virtue of the quantities we eat (about 125 pounds per person each year), potatoes are the leading source of vitamin C in the American diet.

Stuffed Potatoes

If you've always thought of baked potatoes as a side dish, be assured that, combined with low-fat dairy products, they can be a satisfying, protein-rich main dish. Accompany with a colorful assortment of side vegetables and a salad. Maybe you'll discover, as I have, that potatoes can be the best part of a meal!

4 large russet potatoes
1 tablespoon butter or soy margarine,
    at room temperature
1/2 cup buttermilk
1/4 cup low-fat cottage cheese
2 tablespoons minced fresh chives
1/4 teaspoon salt

Dash of freshly ground black pepper
Dash of freshly grated nutmeg
Dash of paprika
1/4 cup (1 oz.) shredded Cheddar
    cheese
Parsley sprigs, for garnish

Preheat oven to 350F (175C). Scrub potatoes and pierce in several places with a sharp knife. (This prevents a buildup of steam, which could cause the potatoes to explode.) Bake until tender, about 45 minutes.

Remove from oven, cut each in half lengthwise and let cool slightly. Scoop out the potato centers, leaving a firm, 1/4-inch shell. Mash the potatoes with the butter.

Place the buttermilk and cottage cheese in a small bowl and microwave on HIGH 1 minute. Stir buttermilk mixture, chives, salt, pepper and nutmeg gently into mashed potatoes. They can be a little lumpy. Pack the potato mixture back into the shells. Dust with paprika and top each half with a little Cheddar cheese. Bake until potatoes are heated through and cheese is melted, about 15 minutes. Garnish with parsley.

Makes 4 servings.

Chicago-Style Deep Dish Pizza

This style of pizza is a deep dish pie featuring a layer of cheese beneath the tomato topping, assuring that the crust won't become soggy. Friends rave about this one!

Crust	PIZZA CRUST
1 (28-oz.) can Italian plum tomatoes, drained	1 cup lukewarm water
4 garlic cloves, minced	1 teaspoon sugar
2 tablespoons thinly sliced fresh basil leaves	1 tablespoon active dry yeast
2 cups (8-oz.) shredded mozzarella cheese	2 cups unbleached all-purpose flour
Freshly grated Parmesan cheese	1/2 cup whole-wheat flour
	3/4 cup stone-ground yellow cornmeal
	1 teaspoon salt
	3 tablespoons plus 1 teaspoon olive oil

Prepare crust.

Meanwhile, preheat oven to 475F (245C). Seed and finely chop the plum tomatoes, removing any hard, woody cores. Drain tomatoes a second time after chopping. You want the topping dry, so the crust doesn't absorb liquid. Combine the chopped tomatoes, garlic and basil in a medium bowl.

Prick the crust liberally across the bottom and up the sides with the tines of a fork. Bake crust 5 minutes. Remove and brush with remaining teaspoon of oil.

Sprinkle the mozzarella cheese over the bottom of the crust. Spoon on tomato mixture. Sprinkle with Parmesan cheese. Place in the bottom third of the oven and bake 30 minutes, until the crust is golden brown and cooked through, and the cheese is bubbly. Slice and serve immediately.

Makes 4 servings.

Pizza Crust

Combine the water and sugar in a warmed small bowl, and stir until sugar is dissolved. Sprinkle yeast over the water and let stand 5 minutes, or until frothy.

Place the flours, cornmeal and salt in a food processor and process until combined.

Add the yeast mixture and 3 tablespoons of the oil and process until the dough holds together and forms a smooth ball.

Remove the dough from the food processor and place in a large, lightly oiled bowl. Cover with plastic wrap and let stand in a warm place about 30 minutes, or until it is light. Punch dough into a 15-inch, deep-dish pizza pan (See Note, below.) Cover the dough with plastic wrap and let it rise in a warm place 20 minutes.

Education

One of the things to remember when making a quick homemade pizza crust is that yeast likes to be warm. If it gets a chill, it won't work as hard to raise the dough.

While lukewarm water from the tap (it should feel quite warm, but not too hot to the touch) is sufficient to get the yeast growing, they'll slow down if the mixture is put into a cold bowl.

To warm the bowl, you can: (1) place it in a warm oven; (2) place a cup of water in the bowl, then heat it in the microwave; or (3) rinse the bowl with hot tap water or boiling water. Likewise the large bowl in which the dough will rise should provide a warm resting place. Once the dough has risen, been punched down and pressed into the pan, let it rise in a warm place to complete the growth of the yeast.

Note I've also used a 12-inch shallow tart pan instead of the pizza pan. This makes a thicker crust, which is quite nice, too. You can experiment with the size and shape of your baking pans and not go too far wrong, but do use metal, which will conduct heat more quickly than ceramic or glass.

Tamale Pie

Wonderful, comforting food and a gorgeous presentation! The spicy bean filling is a tasty counterpoint to the cornmeal crust and needs only a salad to make a complete meal.

3 ears fresh corn, with husks, or
    1 cup frozen whole-kernel corn
3/4 cup stone-ground cornmeal
4 cups water
1/2 teaspoon salt
2 tablespoons olive oil
1 cup chopped onion
1/2 cup diced green bell pepper
1 jalapeño chile, seeded and minced
1 tablespoon minced fresh garlic
2 tablespoons chili powder
1 teaspoon ground coriander
1 teaspoon ground cumin
1 2/3 cups cooked (or 15-oz. can)
    kidney beans, drained
1 2/3 cups cooked (or 15-oz. can)
    pinto beans, drained
1 (8-oz.) can tomato sauce
1/2 cup sliced ripe olives
1/4 cup (1 oz.) shredded Cheddar
    cheese

Preheat oven to 350F (175C). Lightly oil a 10-inch glass pie plate.

Microwave the corn in husks 9 minutes on HIGH. Set aside to cool. When corn is cool enough to handle, remove husks and silk, cut off kernels and scrape cobs with a knife to remove remaining corn. Set aside.

Combine the cornmeal, water and salt in a large microwave-safe bowl. Stir well. Microwave, uncovered, on HIGH 6 minutes. Stir well again and cover bowl with waxed paper. Microwave on HIGH 6 minutes more. Stir and set aside, covered.

Heat the oil in a large, nonstick skillet over medium heat. Add the onion and bell pepper and cook, stirring frequently, until the onion is translucent, about 3 to 4 minutes. Add chile, garlic, chili powder, coriander and cumin and stir 1 minute. Stir in kidney and pinto beans, tomato sauce, olives and fresh or frozen corn, and bring to a boil. Cook 1 minute, stirring occasionally, then remove from heat.

To assemble pie, spread half of the cornmeal mush across the bottom and up the sides of a 10-inch glass pie plate. Spoon the bean mixture over top and spread evenly. Top with remaining cornmeal mixture and spread evenly across surface, to edges of pie plate. Sprinkle with cheese.

Set pie plate on a baking sheet to catch any runover. Bake 50 to 55 minutes, or until browned and bubbly. Bring whole pie to the table and spoon portions onto individual plates.

Makes about 6 servings.

Education

Although corn is high in protein, it must be combined with beans to balance its deficiencies. A traditional dish like Tamale Pie combines beans and cornmeal, and, in this case, fresh or frozen corn. The result is a complete protein dish without cholesterol and with a minimum of fat.

Scalloped Vegetable & Cheese Casserole

A variety of vegetables, layered and baked, yields a nutritious, tummy-satisfying meal.

2 russet potatoes
1 stalk broccoli
1 onion
2 carrots
1 medium zucchini
2 eggs
1 1/2 cups low-fat milk

1 garlic clove, minced
1/2 teaspoon salt
1/8 teaspoon freshly ground black
    pepper
2/3 cup (about 2 1/2 oz.) shredded
    Cheddar cheese

Preheat oven to 350F (175C). Lightly oil a shallow 13 x 9-inch casserole dish.

Peel and thinly slice potatoes. Layer across the bottom of prepared casserole dish. Peel broccoli stem and thinly slice. Chop broccoli top. Layer over the potatoes. Peel onion, cut in half lengthwise and thinly slice. Separate slices and arrange over broccoli. Peel and thinly slice carrots on the diagonal. Arrange over onion. Thinly slice zucchini on the diagonal and arrange over carrots.

Beat together eggs, milk, garlic, salt and pepper in a medium bowl. Pour over vegetables. Cover with foil and bake 60 minutes, or until the vegetables are nearly tender. Remove foil and sprinkle casserole with cheese. Bake 10 to 15 minutes or until casserole top is lightly browned.

Makes 4 servings.

Cheese-Crusted Onion Dumplings

Tender onions nestle in biscuit dough that is delicately flavored with cheese for a tempting main dish. The touch of fennel in the dough gives a wonderful aroma to this dish as it is removed from the oven.

4 medium onions
1 cup whole-wheat flour
2 teaspoons baking powder
1/2 teaspoon salt
1/2 teaspoon ground yellow mustard
1/4 teaspoon ground fennel seeds
3 tablespoons soy margarine or
    butter
1/2 cup (2 oz.) shredded sharp
    Cheddar cheese
About 2/3 cup low-fat milk

4 teaspoons minced fresh herbs or
    sprigs of parsley, tarragon,
    marjoram or cilantro
1/4 cup tomato or eight-vegetable
    juice
2 tablespoons water
1 tablespoon olive oil
1 tablespoon soy sauce
1/2 garlic clove, minced
1/2 teaspoon cornstarch

Preheat oven to 350F (175C). Lightly oil bottom of a 9 x 9-inch baking pan.

Bring a large saucepan of water to a boil. Trim and peel onions. Add onions to pan and cook, uncovered, about 35 to 40 minutes, or until the onions are tender but still retain their shape.

Combine the flour, baking powder, salt, mustard and fennel in a large bowl. Cut in the margarine and cheese with a pastry blender (or in a food processor) until the mixture resembles coarse crumbs. Stir in milk, just until dough holds together.

Divide the dough into 4 pieces. Roll out each piece on a lightly floured surface to about a 1/2-inch-thick square that is large enough to hold an onion.

Drain onions well. Place each onion on a square of dough. Sprinkle with minced herbs or top with 1 or 2 herb sprigs. Fold dough up and pinch the corners together over the tops of the onions. Place in prepared baking pan. Bake 35 to 40 minutes, or until crust is browned and cooked throughout.

While dumplings are baking, prepare sauce. Combine juice, water, oil, soy sauce, garlic and cornstarch in a small saucepan and stir until cornstarch is dissolved. Cook, stirring, over medium heat until mixture comes to a boil. Cook, stirring, 1 minute more, or until mixture is slightly thickened and the garlic has mellowed. Spoon hot sauce over dumplings just before serving.

Makes 4 servings.

Variation

Onion Dumplings can be served at room temperature or packed for a picnic. In this case, omit the sauce.

Education

If you're accustomed to serving meats, gravy may be a familiar staple on the dinner table. Gravy adds moisture and flavor to the foods it garnishes, but often at the expense of high fat. The sauce, above, is thick and flavorful, without a lot of fat. It will highlight the dumplings without eclipsing their delicate flavor.

Baked Macaroni Noodle Casserole

Macaroni and cheese wins a slot among many families' favorites. Economical and filling, this traditional dish provides the complementary proteins of dairy (cheese) and grains (wheat noodles). Onion, green bell pepper and tomato sauce lifts this version out of the ordinary.

2 cups macaroni	1/4 cup finely diced celery
1 1/2 cups (6 oz.) shredded sharp Cheddar cheese	1/4 cup finely diced green bell pepper
1 cup fresh whole-grain bread crumbs	1/4 cup grated onion
1 egg, beaten	Dash of freshly grated nutmeg
1 cup tomato sauce	Dash of paprika
1 cup low-fat milk	Freshly grated Parmesan cheese, for garnish

Preheat oven to 350F (175C). Lightly oil a deep casserole dish.

Bring a large saucepan of water to a boil. Stir in the macaroni and cook until quite firm-tender, about 5 minutes. Drain and rinse briefly under cold running water.

Layer 1/3 of the macaroni, 1/3 of the cheese and 1/3 of the bread crumbs in prepared dish. Repeat with remaining macaroni, cheese and bread crumbs, making 2 more layers.

Combine the egg, tomato sauce, milk, celery, bell pepper, onion, nutmeg and paprika. Pour tomato mixture over the macaroni, using a spoon to help the liquid reach the bottom of the baking dish, if necessary. Top casserole with Parmesan cheese.

Bake casserole 45 minutes, or until browned at the edges.

Makes 6 servings.

Education

If you're trying to introduce more whole grains to the table, and meeting some resistance, add whole-grain bread crumbs or even pasta to dishes like the one here. The hearty flavor of a casserole can be complemented by the full grains, allowing everyone to benefit from more fiber and complete nutrition.

Late-Summer Garden Casserole

Rich layers of summer's bounty are crowned with a light layer of cheese. Serve with rice and a tomato salad.

2 medium zucchini
1 medium eggplant
1 large red bell pepper
2 large onions
12 to 16 basil leaves
1 teaspoon minced fresh thyme
    leaves or 1/4 teaspoon dried
    thyme
2 tablespoons olive oil
2 garlic cloves, minced

1 shallot, minced
1/4 cup tomato juice
1/4 teaspoon salt
Dash of freshly ground black pepper
1 cup (4 oz.) shredded mozzarella
    cheese
2 tablespoons freshly grated
    Parmesan or Asiago cheese
2 tablespoons minced fresh parsley

Preheat oven to 375F (190C). Lightly oil a 13 x 9-inch nonstick casserole dish.

Remove stem and blossom ends from zucchini and cut crosswise into 1/4-inch-thick slices. Trim and peel eggplant and cut crosswise into 1/2-inch-thick slices. Remove core and seeds from bell pepper and cut into 1/4-inch-thick rings. Peel onions and cut into thin slices. Layer the vegetables in prepared casserole dish, alternating with basil and thyme.

Heat oil in a small pan over medium-low heat. Add the minced garlic and shallot and cook, stirring constantly, 1 minute, or until shallot and garlic begin to soften. Stir the tomato juice, salt and black pepper into the pan. Spoon tomato juice mixture over the vegetables in the casserole dish.

Bake the casserole 35 to 40 minutes, or until the vegetables are firm-tender. Combine cheeses with parsley and sprinkle over the casserole. Bake 10 to 15 minutes, or until the cheeses are melted and the vegetables are tender.

Makes 4 servings.

Cheese-Baked Enchilada Casserole

Bubbly cheese tops layer upon layer of tortillas with spicy bean and vegetable filling and tangy salsa. Add a salad for a simply stupendous meal!

1 tablespoon olive oil
1 large onion, chopped
1 tablespoon minced garlic
1 small jalapeño chile, seeded
    and minced
1 tablespoon chili powder
2 teaspoons ground cumin
1 teaspoon ground coriander
1/2 teaspoon dried oregano,
    crumbled
1/2 teaspoon salt
2 cups finely diced zucchini
1 green bell pepper, finely diced
1/4 cup water
1 2/3 cups cooked (or 15-oz. can)
    red kidney beans, drained

1 cup fresh tomatoes, diced, peeled
    and seeded
1 cup fresh or frozen whole-kernel
    corn
1 tablespoon minced fresh cilantro
4 cups Garlic-Lovers' Salsa Cruda
    (page 3) or 3 (14 1/2-oz.)
    cans Mexican-style stewed
    tomatoes, blended until smooth
12 corn tortillas
2 cups (8 oz.) shredded sharp
    Cheddar cheese
1 cup (4 oz.) shredded Monterey
    Jack cheese

Preheat oven to 350F (175C).

Heat the oil in a nonstick Dutch oven over medium heat. Add onion and cook, stirring frequently until the onion is translucent, about 3 to 4 minutes. Add the garlic and chile and cook, stirring, 1 minute. Add chili powder, cumin, coriander, oregano and salt and stir until well combined.

Stir in zucchini, bell pepper and water. Cover and steam 3 to 4 minutes, or until the zucchini is just beginning to get tender. Stir in beans, tomatoes and corn and heat through. Remove from heat and stir in cilantro.

To assemble casserole, spread 1/2 cup salsa or blended tomatoes over the bottom of a 13 x 9-inch baking dish. Arrange 4 of the tortillas over the sauce. Top with half of the bean mixture, spreading evenly.

Combine the cheeses. Sprinkle 1 cup of cheese over the beans. Spoon on about 1/3 of the remaining salsa. Top with 4 tortillas and the remaining bean mixture. Sprinkle with 1 cup of cheese and half the remaining salsa.

Arrange remaining 4 tortillas over top and spoon on remaining salsa. Top with remaining cheese.

Bake 40 minutes, or until casserole is bubbly and vegetables are tender. Remove from oven and let stand 10 minutes before cutting.

Makes 8 servings.

Eggplant Mozzarella

An attractive main dish, served with polenta, couscous or pasta and tomato sauce. Add a green salad and enjoy . . .

1 medium eggplant
1/2 cup unbleached all-purpose flour
1 egg, beaten
1/4 cup low-fat milk
2 tablespoons stone-ground yellow
    cornmeal

1 cup (4 oz.) shredded mozzarella
    cheese
1/2 cup tomato sauce

Preheat oven to 350F (175C). Lightly oil a large nonstick shallow baking pan.

Cut the eggplant crosswise into 8 equal slices. Combine egg, milk and cornmeal in a shallow bowl. Dip slices first in flour, then in batter, stirring batter after each dip.

Arrange eggplant slices on prepared baking pan. Bake 45 to 50 minutes, or until eggplant is tender. Top with cheese and spoon some of the sauce across each slice. Bake about 10 minutes, or until cheese is lightly melted.

Makes 4 servings.

Education

Eggplant can absorb more oil in cooking than any other vegetable. Deep-frying eggplant can add up to 700 calories per serving in fat alone. For a healthy take on cooking eggplant, bake as above, steam or stew. If you're broiling or grilling eggplant, brush lightly with oil and leave it at that.

Vegetable Pot Pie

A tender, whole-wheat crust tops a subtle mix of savory vegetables in the Pennsylvania Dutch tradition.

1 large onion, halved and thinly
    sliced crosswise
2 tablespoons butter or soy margarine
2 carrots, diced
1/4 yellow summer squash, diced
2 tablespoons whole-wheat flour
1 2/3 cups vegetable broth
3/4 cup low-fat milk
1/4 teaspoon salt
Dash of freshly ground black pepper
3/4 cup frozen green peas
1 tablespoon minced fresh parsley
1/2 teaspoon minced fresh marjoram
    or 1/8 teaspoon dried marjoram
Dash of freshly grated nutmeg

WHOLE-WHEAT CRUST
1 cup whole-wheat flour
1 cup unbleached all-purpose flour
1/2 teaspoon salt
8 tablespoons soy margarine or
    butter, chilled
5 to 6 tablespoons cold water

On a floured piece of waxed paper, roll out the crust to fit a 13 x 9-inch baking dish with 1/2 inch extra dough all around. (Rub a damp cloth on the kitchen counter to hold the waxed paper still as you roll.) Cut a design in the surface of the crust to allow steam to escape when the pie is baking. Set crust aside on waxed paper.

Meanwhile, preheat oven to 375F (190C). Lightly oil a 13 x 9-inch baking dish.

Combine the onion and the butter in a large microwave-safe bowl. Microwave, uncovered, on HIGH 2 minutes. Stir in carrots. Microwave, uncovered, on HIGH 2 minutes. Add squash and stir. Microwave, uncovered, on HIGH 3 minutes.

Stir in flour. Stir in broth, milk, salt and pepper. Microwave, uncovered, on HIGH 1 minute and stir. Continue cooking in 1-minute periods and stirring until mixture is slightly thickened. Stir in peas, parsley, marjoram and nutmeg.

Transfer vegetable mixture to prepared baking dish. Invert the crust over filling and gently remove waxed paper. Crimp to form a decorative edge, or double over and press gently with the tines of a fork.

Bake pie 40 to 45 minutes, or until crust is golden brown.

Makes 6 servings.

Whole-Wheat Crust

Combine the flours and salt in a large bowl. Cut in the butter or margarine with a pastry cutter, or work in with your fingers until mixture is crumbly and well-blended. Sprinkle with 4 tablespoons of the water. Quickly work the dough until it is just soft enough to hold together, adding additional water as necessary. Cover bowl with plastic wrap and refrigerate dough 30 minutes. (This relaxes the gluten so the crust stays tender.) Wrapped well so that it does not dry out, the dough can be refrigerated up to a week. Bring back to room temperature to roll out.

Variation

In winter substitute 3/4 cup diced, cooked winter squash or potato for yellow summer squash.

Baked Mushroom-Stuffed Tomatoes

Use the really large garden-ripe tomatoes for best results. Serve with lightly steamed green beans and ears of corn.

4 large tomatoes
1 teaspoon salt
1/4 cup couscous
1/2 cup boiling water
1 tablespoon fresh lemon juice
1 tablespoon plus 2 teaspoons
    olive oil
2 cups sliced mushrooms
2 tablespoons thinly sliced green
    onions

2 tablespoons finely diced green
    bell pepper
3/4 cup fresh whole-wheat bread
    crumbs
1 tablespoon finely chopped walnuts
4 basil leaves, thinly sliced
1/3 cup (about 1 1/2 oz.) shredded
    mozzarella cheese

Preheat oven to 400F (205C).

Cut a thin slice from the top of each tomato. Remove any of the hard, woody core. Scoop out seeds and pulp, discarding seeds and setting pulp aside in a sieve to drain. Leave enough of the tomato shells so they hold their shape. Sprinkle the insides of the tomatoes with salt and turn them upside-down on paper towels to drain 15 minutes.

Combine the couscous with boiling water, lemon juice and the 2 teaspoons oil in a small bowl. Cover and let stand 5 to 10 minutes, until the water is absorbed.

Meanwhile, heat the remaining 1 tablespoon oil in a nonstick skillet over medium heat. Add the mushrooms, green onions and bell pepper and cook, stirring occasionally, until lightly browned, about 5 to 6 minutes.

Squeeze any remaining liquid from the tomato pulp. Chop pulp fine and toss lightly with cooked mushroom mixture, bread crumbs, walnuts, basil and couscous.

Lightly fill each tomato with mushroom mixture. Place in a shallow 9 x 9-inch baking dish. Cover loosely with foil. Bake 20 to 25 minutes, or until tomatoes are tender. Remove foil. Top with shredded cheese and return to oven just until cheese melts.

Makes 4 servings.

Education

Mushroom varieties are sprouting up in supermarkets everywhere! Try shiitake, crimini or porcini mushrooms in place of the usual button mushrooms. You can afford to explore to your heart's content, for not only are mushrooms low in calories, but they are high in potassium and low in sodium—the perfect combination for helping to control blood pressure.

Noodle-Spinach Bake with Three Cheeses

Feta and Armenian string cheese add a Middle Eastern twist to a simple casserole. Serve with cooked carrots and a tomato salad.

1 1/2 cups curly egg noodles
1 pound spinach, stems removed
2 eggs
1/4 cup whole-wheat flour
1 cup low-fat cottage cheese
1/2 cup diced Armenian string cheese
1/4 cup crumbled feta cheese
1 tablespoon minced fresh parsley
1 garlic clove, minced

Preheat oven to 350F (175C). Lightly oil a 13 x 9-inch shallow baking dish.

Cook the noodles in a large saucepan of boiling water about 6 to 7 minutes, or just until barely tender. Drain and let cool slightly.

Meanwhile, wash spinach well. Cook in a large nonstick skillet over medium heat, with just the water that clings to the leaves, stirring constantly, just until wilted. Place spinach in a sieve and press out any liquid. Chop the spinach and set aside.

In a large bowl, beat the eggs and combine with the flour. Stir in cottage cheese, string cheese, feta, parsley and garlic. Add noodles and spinach and stir until combined.

Spoon the mixture into prepared baking dish. Cover with foil. Bake 35 to 40 minutes, or until mixture is set.

Makes 4 servings.

Monterey Mushroom & Potato Strata

"Strata" refers to the layers of vegetables enhanced with cheeses—simple, but very good.

3 large red potatoes (about 1 pound)
1 (15-oz.) container low-fat ricotta
    cheese
2 cups (8-oz.) shredded Monterey
    Jack cheese
1/4 cup minced fresh parsley

2 tablespoons thinly sliced basil
    leaves
1 garlic clove, minced
2 cups sliced button mushrooms
1 onion, halved and thinly sliced

Preheat oven to 400F (205C). Lightly oil a 14 x 11-inch shallow baking dish.

Peel and slice potatoes. Arrange in a layer in the bottom of prepared baking dish. Combine the ricotta, half the Monterey Jack cheese, parsley, basil and garlic in a medium bowl, stirring until well-combined. Spread the cheese mixture over the potatoes.

Combine the mushrooms and onion in a medium bowl. Arrange over the cheese mixture. Top with remaining Monterey Jack.

Bake 35 to 40 minutes, or until top is browned and potatoes are tender. Let stand about 5 minutes before serving.

Makes 6 servings.

Education

Fresh mushrooms have a decided nutritional edge over canned varieties. Cook them yourself, and you'll have fifteen times the riboflavin, three times the niacin and potassium and twice the iron of an equal amount of canned mushrooms. Canned mushrooms are also likely to contain added, and unneeded, sodium.

Diavolo Golden-Stuffed Peppers

Golden yellow peppers feature a corn stuffing with fresh herbs and a hint of hot chile to "light" up your table!

3 large yellow bell peppers
1 tablespoon olive oil
1 onion, chopped
1/2 teaspoon minced jalapeño chile
2 garlic cloves, minced
1/2 teaspoon ground coriander
1/4 teaspoon ground cumin
2 large tomatoes, peeled, seeded and finely chopped

2 cups whole-kernel corn, thawed if frozen
1 cup fresh whole-wheat bread crumbs
1/2 cup (2 oz.) shredded sharp Cheddar cheese
2 tablespoons minced fresh parsley
1/2 teaspoon salt
Dash of freshly ground black pepper

Preheat oven to 350F (175C).

Halve peppers lengthwise, remove seeds and carefully remove hard core. Set on a microwave-safe plate and microwave on HIGH 2 minutes. Set aside.

Heat the oil in a large, nonstick skillet over medium heat. Add the onion and chile and cook, stirring frequently, until the onion is translucent, about 3 to 4 minutes. Add garlic, coriander and cumin and cook, stirring, 1 minute.

Remove pan from heat. Stir in tomatoes, corn, bread crumbs, cheese, parsley, salt and pepper until well combined. Lightly pack each pepper half with some of the mixture and place in a 9 x 9-inch baking dish. Cover loosely with foil. Bake 30 to 40 minutes, or until the peppers are tender.

Makes 6 servings.

Education

Corn, though high in protein, contains limited amounts of the essential amino acids lysine and tryptophan. By combining a small amount of cheese with the corn in this dish, you can compensate for these low levels and enjoy a protein-rich main dish.

Orzo & Cheese–Stuffed Bell Peppers

Fresh squash and bell peppers make tasty companions on your dinner plate. Use your favorite prepared pasta sauce to add flavor without hassle.

2 tablespoons olive oil
1 small onion, chopped
1/2 cup sliced mushrooms
1/2 cup shredded summer squash
1/2 cup shredded carrots
2 large garlic cloves, minced
1 (15-oz.) container part-skim
    ricotta cheese
1 cup cooked orzo or rice
1 egg, beaten
2 tablespoons minced fresh parsley

1 tablespoon thinly sliced fresh basil
    leaves or 1/4 teaspoon dried
    marjoram
1/2 teaspoon salt
Dash of freshly ground black pepper
4 large red bell peppers, halved
    lengthwise
1 1/2 cups tomato pasta sauce with
    mushrooms
1/2 cup shredded mozzarella cheese

Preheat oven to 350F (175C).

Heat the oil in a large, nonstick skillet over medium heat. Add the onion, mushrooms, squash and carrots and cook, stirring occasionally, until the vegetables are firm-tender, about 6 to 8 minutes. Add garlic and cook, stirring, 1 minute. Set aside.

Combine ricotta, orzo or rice, egg, parsley, basil or marjoram, salt and pepper in a medium bowl. Stir in vegetable mixture.

Remove stems and seeds from pepper halves. Divide the cheese mixture among the pepper halves.

Pour about half of the tomato sauce in the bottom of a 13 x 9-inch baking pan. Arrange stuffed pepper halves in pan and spoon some of the remaining sauce over peppers. Cover with foil and bake 40 to 45 minutes or until filling is hot and peppers are firm-tender.

Uncover and sprinkle peppers with cheese. Bake 10 minutes, or until peppers are tender and cheese is melted.

Makes 4 servings.

Education

Native to the Western Hemisphere, bell peppers became popular nearly all over the world, following their discovery by Spanish explorers. Peppers are surprisingly nutritious: Who would think that these tasty vegetables outshine citrus fruits in vitamin C? Yet green bell peppers contain, ounce-for-ounce, twice the vitamin C of citrus. Sweet red peppers (green peppers that have ripened) are even more nutritious, containing three times the vitamin C of citrus and over ten times the amount of beta carotene found in a green pepper.

Oven-Cooked Dried Beans

Having cooked beans on hand, either refrigerated or frozen, will make meal preparation quicker and easier. One easy method is to oven-bake the beans after soaking overnight. They cook without stirring and without fuss.

For each type of bean, pick over 1 pound of beans and remove any stones or moldy beans. Rinse well. Place in a very large casserole dish and cover well with water. Let stand overnight. (Lentils are quicker cooking and need no pre-soaking.)

Drain soaked beans, return to casserole dish and cover again generously with water. Cover casserole and bake at 350F (175C) according to the times below. (Times can vary according to how long the beans have been stored, growing conditions, etc.)

TYPE OF BEAN	COOKING TIME
BLACK BEANS	1 TO 1 1/4 HOURS
CHICKPEAS	3 TO 3 1/2 HOURS
LENTILS (NO SOAKING)	45 MINUTES TO 1 HOUR
NAVY BEANS	1 1/2 TO 2 HOURS
PINTO BEANS	2 1/2 TO 3 HOURS
RED OR WHITE KIDNEY BEANS	2 TO 2 1/2 HOURS
SOYBEANS	3 TO 3 1/2 HOURS

Company's Coming Cheese-Stuffed Shells

Spinach and three cheeses flavor the stuffed shells, which can be prepared with your favorite pasta sauce if time is limited. Serve with a big tossed salad, and invite friends.

3/4 pound jumbo shells
1 cup packed fresh spinach leaves
3 eggs
1 (15-oz.) container fat-free ricotta cheese
1 1/2 cups (6 oz.) shredded mozzarella cheese
3/4 cup freshly grated Parmesan cheese

1 tablespoon minced fresh parsley
1/2 teaspoon salt
Dash of freshly ground black pepper
4 cups homemade tomato pasta sauce or 1 (32-oz.) jar tomato pasta sauce

Preheat oven to 350F (175C). Lightly oil a 13 x 9-inch shallow casserole dish.

Bring a large pot of water to a boil. Add jumbo shells a few at a time, stirring after each addition, so the shells stay separate while cooking. Return to a boil and cook, uncovered, 8 to 10 minutes, or just until firm-tender. Do not overcook. Drain and cool in a single layer on a sheet of waxed paper to prevent shells from sticking together.

Wash spinach well in lukewarm water to remove grit. Cook in a medium, nonstick skillet in just the water that clings to the leaves, tossing just until wilted. Wrap in a paper towel and press out moisture. Finely chop cooked spinach.

Beat the eggs in a large bowl. Add spinach and stir until combined. Add mozzarella cheese, 1/2 cup of the Parmesan cheese, parsley, salt and pepper and stir until all ingredients are well combined. Fill each jumbo shell with 1 1/2 to 2 tablespoons of the spinach mixture.

Pour 1 cup of the pasta sauce in the bottom of prepared casserole dish. Arrange the shells in one layer. Spoon over the remaining sauce. Cover loosely with foil. Bake 30 minutes. Remove foil and sprinkle with the remaining 1/4 cup Parmesan cheese. Bake, uncovered, 5 minutes, or until cheese is melted.

Makes 8 servings.

Education

Try the new fat-free dairy products on the market to continue enjoying old favorites while cutting back on fat and calories. Substituted in whole or part for higher-fat counterparts, the new fat-free ingredients may cause little or no change in flavor in familiar dishes, but can support better health.

Tortilla Flats Green-Chili Casserole

Just pop this easy-to-make casserole into the oven and go about your business until it's table-ready! Heat the salsa topping in the microwave, and then a quick salad is all you'll need to complete the meal.

1 tablespoon olive oil
1 large onion, chopped
3 garlic cloves, minced
1/2 teaspoon ground coriander
1/2 teaspoon ground cumin
1/4 teaspoon chili powder
1/8 teaspoon dried oregano
12 corn tortillas
3/4 cup (3 oz.) shredded sharp
    Cheddar cheese

3/4 cup (3 oz.) shredded Monterey
    Jack cheese
1 (4 1/2-oz.) can mild chopped
    green chiles, drained
3 eggs
2 egg whites
2 1/2 cups buttermilk
1/2 teaspoon salt

Topping
1 cup salsa
1 cup tomato sauce

Preheat oven to 375F (190C). Lightly oil a 13 x 9-inch baking dish.

Heat the oil in a medium skillet over medium heat. Add onion and cook, stirring frequently, until the onion is translucent, about 3 to 4 minutes. Add garlic, coriander, cumin, chili powder and oregano and cook, stirring, 1 minute. Remove from heat.

Tear the tortillas into 1-inch pieces. Arrange half the tortilla pieces across the bottom of prepared baking dish.

Combine the cheeses in a medium bowl. Sprinkle half of the cheese mixture over the tortillas and top with half the chopped chiles. Spread cooked onion over all. Repeat with a layer of the remaining tortilla pieces, cheese and chiles.

Beat the eggs, egg whites, buttermilk and salt in a medium bowl. Pour over casserole. Bake about 40 to 45 minutes, or until mixture is set and browned on top.

Place salsa and tomato sauce in a small microwave-safe serving bowl. Microwave on HIGH 2 minutes or until hot. Serve on the side.

Makes 6 servings.

Education

Salsa is a great low-fat topping for a variety of dishes. Salsa can also be stirred into soups to add zip. Since salsa is made up of vegetables high in vitamins C and A, you'll be adding nutrition, not just flavor, to your meals.

Pepper & Olive Meatless Loaf

A centerpiece for an almost traditional dinner with mashed potatoes and green peas.

1 cup brown lentils	1 teaspoon ground coriander
1 bay leaf	1/2 teaspoon ground cumin
3 cups water	1/4 teaspoon dried oregano
1 large garlic clove, peeled	1/4 cup minced ripe olives
2 whole cloves	1/2 cup walnuts
1 onion, finely chopped	1/2 cup fresh bread crumbs
1 green bell pepper, finely diced	1/2 cup low-fat milk
2 teaspoons olive oil	1 egg
1 carrot, coarsely shredded	1 tablespoon soy sauce
1 garlic clove, minced	1/4 cup ketchup

Bring the lentils, bay leaf and water to a boil in a large saucepan over medium-high heat. Stick the garlic clove with the whole cloves and add to pan. Reduce heat, cover and simmer 20 minutes, or until lentils are firm-tender. Remove bay leaf and garlic with cloves. Let cool slightly and drain.

Preheat oven to 350F (175C). Lightly oil a 9-inch-square shallow casserole dish.

Combine the onion, bell pepper and oil in a large microwave-safe bowl and microwave, uncovered, on HIGH 2 minutes. Stir in carrot, minced garlic, coriander, cumin and oregano. Microwave, uncovered, on HIGH 1 minute. Stir in olives.

Place the lentils, walnuts, crumbs, milk, egg and soy sauce in a food processor and process until ingredients are coarsely chopped. Add the onion mixture and stir until combined.

Spoon lentil mixture into prepared dish. Spread top with ketchup. Bake 30 minutes or until set.

Makes 6 servings.

Cornmeal-Crusted Two-Bean Chili

Baked chili in a casserole with a hearty cornmeal topping needs only an accompanying salad to make it a meal. The chili is jam-packed with tasty, nutritious vegetables.

2 tablespoons olive oil
1 large onion, chopped
1 large carrot, diced
1/2 red bell pepper, diced
1 tablespoon minced garlic
1 tablespoon chili powder
2 teaspoons ground coriander
2 teaspoons ground cumin
1/2 teaspoon ground yellow
    mustard
1/8 teaspoon ground cayenne
    (optional)
1 (28-oz.) can whole tomatoes,
    chopped, with juice
1 2/3 cups cooked (or 15-oz. can)
    red kidney beans, drained
1 2/3 cups cooked (or 15-oz. can)
    pinto, pink or black beans,
    drained

1 tablespoon minced fresh cilantro
    or parsley
1/4 teaspoon freshly ground black
    pepper
1 cup unbleached all-purpose flour
3/4 cup stone-ground yellow
    cornmeal
1 1/2 teaspoons baking powder
1/2 teaspoon baking soda
1/2 teaspoon salt
2 eggs, beaten
1 cup buttermilk
1/2 cup (2 oz.) shredded Cheddar
    cheese
1/4 cup fat-free sour cream

Preheat oven to 400F (205C). Lightly oil a 13 x 9-inch baking pan.

Heat the oil in a large, nonstick skillet over medium-high heat. Add the onion, carrot and bell pepper and cook, stirring frequently, until the onion is translucent, about 3 to 4 minutes. Add the garlic, chili powder, coriander, cumin, mustard and cayenne, if desired, and cook, stirring, 1 minute.

Add chopped tomatoes with juice, beans, cilantro or parsley and black pepper to skillet. Partially cover pan and bring to a boil. Reduce heat and simmer until vegetables are tender and chili is thickened, about 5 minutes.

Combine flour, cornmeal, baking powder, soda and salt in a medium bowl. In another bowl, combine the eggs, buttermilk, cheese and sour cream. Stir egg mixture into dry ingredients just until combined.

To assemble casserole, spoon the chili into prepared baking pan. Spoon cornmeal topping over chili, and spread evenly with a rubber spatula. Place casserole dish on a baking sheet. Bake 25 to 30 minutes, or until topping is golden brown and firm. Remove from oven and let stand 5 minutes before serving.

Makes 6 servings.

Your Own Fabulous Pizza

While prepared crusts and pizza "kits" are convenience itself, they can't compare to a crust you make yourself. It isn't even that hard, and the results are worth it! (If you're not convinced, frozen pizza or bread dough can be shaped and baked, as below.)

Pizza Crust (page 74)
2 cups (8 oz.) shredded mozzarella
    cheese
1/2 yellow or green bell pepper,
    thinly sliced

4 to 5 plum tomatoes, peeled and
    thinly sliced
2 tablespoons freshly grated
    Parmesan cheese

Prepare dough and let rise as directed. Preheat oven to 450F (225C).

Preheat a baking sheet in the oven. Punch down and roll out dough on a floured surface to a circle 15 to 16 inches across. Fold under about 1/4 inch all around the edge of the crust. Place on preheated baking sheet.

Sprinkle crust with mozzarella cheese, then arrange pepper strips and tomato slices over top. Place on bottom rack in the oven and bake about 15 minutes, or until crust is golden brown. Remove from oven and sprinkle with Parmesan cheese.

Makes 4 servings.

Variations

Onion & Pesto Pizza Substitute 1 cup thinly sliced onion for bell pepper. Combine onion and 1 teaspoon olive oil in a medium microwave-safe bowl and microwave on HIGH 2 minutes. Toss onion with 2 tablespoons Pesto Genovese (page 61). Top mozzarella cheese with tomatoes, as above, then top with onion mixture. Bake and sprinkle with Parmesan cheese, as above.

If you don't have garden-ripe, flavorful plum tomatoes, drain, seed and chop Italian plum tomatoes from 1 (28-oz.) can. Combine with 2 garlic cloves, minced, and 2 tablespoons minced fresh basil or parsley. Use in place of fresh tomato slices.

Vegetarian Paella

A combination of enticing flavors is what characterizes traditional paella. Here nuts and chickpeas replace fish and chicken, adding appeal to richly flavored saffron rice— wonderful!

4 1/2 cups vegetable broth
1/4 teaspoon saffron threads
1 teaspoon ground coriander
1 teaspoon paprika
1/2 teaspoon salt
2 tablespoons olive oil
1 large onion, chopped
1/3 cup chopped macadamia nuts
    or slivered almonds
1 garlic clove, minced
2 cups long-grain white rice

1 cup peeled, seeded and diced
    tomatoes
1 (8-oz.) package frozen Italian
    green beans, thawed and
    drained
1 (9-oz.) package frozen artichoke
    hearts, thawed and drained
1 cup cooked chickpeas
1 cup frozen green peas
1/4 cup minced fresh parsley

Preheat oven to 350F (175C).

Bring broth, saffron, coriander, paprika and salt to a boil in a large saucepan over medium-high heat. Meanwhile, heat the oil in a nonstick ovenproof Dutch oven over medium heat. Add the onion and macadamias or almonds and cook, stirring frequently, until the onion begins to turn golden, about 5 to 6 minutes. Add garlic and rice and stir until combined.

Pour boiling broth over rice and add tomatoes, green beans, artichoke hearts and chickpeas. Stir until combined. Bring to a boil.

Cover Dutch oven and bake 30 minutes, or until the liquid is absorbed and rice is tender. Stir in peas and parsley. Cover and bake another 10 minutes.

Makes 6 servings.

Zucchini Moussaka

Layers of zucchini, succulent beans and a creamy sauce create a classic with a vegetarian difference.

1 tablespoon olive oil
2 onions, chopped
1 (6-oz.) can tomato paste
1/4 cup water
1/4 cup minced fresh parsley
1/2 teaspoon ground allspice
1/2 teaspoon ground cinnamon
1/4 teaspoon freshly ground black pepper
3 1/3 cups cooked [or 2 (15-oz.) cans] black beans, drained

2 zucchini (about 2 lbs. total), thinly sliced on the diagonal
2 tablespoons butter or soy margarine
2 tablespoons whole-wheat flour
2 cups low-fat milk
1/2 cup freshly grated Parmesan cheese
Dash of freshly grated nutmeg
3 eggs

Preheat oven to 350F (175C). Lightly oil a 13 x 9-inch shallow casserole dish.

Heat the oil in a medium, nonstick skillet over medium heat. Add the onions and cook, stirring occasionally, until the onions begin to brown, about 6 to 7 minutes. Stir in tomato paste, water, parsley, allspice, cinnamon, and pepper and bring to a boil.

Place the beans in a large bowl and mash slightly with a potato masher. Add onion mixture and stir until combined.

Arrange 1/3 of the zucchini slices in a single layer to cover the bottom of prepared casserole dish. Top with half the bean mixture. Arrange 1/3 of the zucchini slices over the beans, and top with remaining bean mixture. Add a layer of the remaining zucchini slices.

Melt butter in a medium saucepan over medium heat. Stir in flour and cook, stirring, 1 minute to cook flour. Stir in milk and cook, stirring, until the sauce thickens. Remove from heat and stir in Parmesan cheese and nutmeg.

Beat eggs in a large bowl. Stir a small amount of sauce into the eggs, then very gradually stir the egg mixture into the sauce. This "tempers" the eggs and keeps them from curdling. Pour sauce over zucchini.

Bake 1 1/4 hours, or until top is nicely browned. Remove from oven and let stand 10 minutes before serving.

Makes 8 servings.

Quesadillas

Beans, cheese, peppers and onions fill flour tortillas that are briefly baked. A selection of low-fat toppings creates a complete, light meal.

1 tablespoon olive oil	1 1/2 cups salsa
1 cup thinly sliced onion	8 flour tortillas
1 cup thinly sliced red bell pepper	1 1/2 cups (6 oz.) shredded sharp Cheddar cheese
2 garlic cloves, minced	
1/2 teaspoon chili powder	3/4 cup shredded zucchini
1/2 teaspoon ground cumin	1/2 cup fat-free sour cream
1 cup cooked beans (kidney, pinto, pink or black beans)	1/2 cup thinly sliced green onions
	2 tablespoons minced fresh cilantro
1/2 teaspoon ground coriander	3/4 cup alfalfa sprouts

Preheat oven to 375F (190C).

Heat the oil in a large, nonstick skillet over medium heat. Add the onion and bell pepper and cook, stirring occasionally, until the vegetables are firm-tender, about 4 to 5 minutes. Add garlic, chili powder and cumin and cook, stirring, 1 minute. Remove from heat.

In a medium saucepan, combine the beans, coriander and 1/2 cup of the salsa. Heat, mashing beans slightly as you stir.

Place 4 of the tortillas on a large baking sheet. Divide and spread beans over each tortilla. Top with the onion mixture, then shredded cheese. Top with the remaining tortillas and press down gently. Spray lightly with cooking oil spray.

Bake 15 minutes, or until cheese is melted. Cut each quesadilla into quarters and serve on individual plates. Top with remaining salsa, shredded zucchini, sour cream, green onions, cilantro and sprouts.

Makes 4 servings.

Biscuit-Topped Corn Casserole

There's something so homey about this dish that if you don't already live on a farm, you might imagine you do, anyway!

1 tablespoon olive oil
1 cup thinly sliced green onions
2 cups chopped broccoli
1 cup shredded carrots
2 cups frozen whole-kernel corn
1 (14 3/4-oz.) can cream-style corn
2/3 cup fat-free sour cream
2 eggs, beaten
1/2 teaspoon dried marjoram
1/4 teaspoon dried thyme

1 1/2 cups unbleached all-purpose
   flour
1/4 cup stone-ground yellow
   cornmeal
1 1/2 teaspoons baking powder
1/2 teaspoon baking soda
1/2 teaspoon salt
4 tablespoons butter or soy
   margarine, chilled
1 1/4 cups buttermilk

Preheat oven to 350F (175C). Lightly oil a 13 x 9-inch nonstick baking pan.

Heat the oil in a large, nonstick skillet over medium heat. Add the green onions and cook, stirring frequently, until they begin to soften, about 2 to 3 minutes. Add the broccoli and carrots and cook, stirring occasionally, until crisp-tender, about 3 to 4 minutes. Add frozen corn and cook, stirring occasionally, just until thawed, about 2 minutes. Remove from heat.

Combine the cream-style corn, sour cream, eggs, marjoram and thyme in a small bowl. Stir into broccoli mixture.

Combine the flour, cornmeal, baking powder, soda and salt in a medium bowl. Cut in the butter with a pastry blender until mixture resembles coarse crumbs. Add buttermilk and stir just until combined.

Pour vegetable mixture into prepared baking pan. Spoon the batter evenly over top. Bake 40 to 45 minutes, or until topping is browned.

Makes 6 servings.

Shepherd's Pie

Traditionally made with meat under a crust of browned mashed potatoes, Shepherd's Pie can also be a satisfying, hearty meatless dish. Portobello mushrooms, onions, carrots and a vegetarian "gravy" flavored with herbs do the trick. Serve with a green vegetable and salad.

6 cups cubed potatoes
1 tablespoon olive oil
2 tablespoons butter or soy
   margarine
1 cup finely diced carrots
1 cup chopped onion
2 garlic cloves
2 cups diced portobello
   mushrooms
2 tablespoons whole-wheat flour

1 1/4 cups vegetable broth
1 tablespoon soy sauce
1/2 teaspoon molasses
1/4 teaspoon freshly ground black
   pepper
1/4 teaspoon dried thyme
1/3 cup low-fat milk
1/3 cup low-fat plain yogurt
1/2 teaspoon salt

Preheat oven to 375F (190C). Lightly oil a deep, 10-inch pie plate.

Cook the potatoes in a large saucepan of boiling water, uncovered, until tender, about 15 to 20 minutes.

Meanwhile, heat the oil and butter in a large, nonstick skillet over medium-high heat. Add the carrots, onion and garlic and cook, stirring occasionally, until the vegetables are firm-tender, about 4 to 5 minutes. Stir in mushrooms and cook, stirring occasionally, until the mushrooms are soft, about 3 to 4 minutes.

Sprinkle flour over the vegetable mixture and cook, stirring, 1 minute. Combine broth, soy sauce and molasses in a 2-cup measure. Add broth mixture, pepper and thyme to vegetable mixture and cook, stirring, until mixture thickens. Remove from heat.

Drain potatoes and return to pan. Combine milk and yogurt in a small glass cup and microwave on HIGH 1 minute to warm. Mash potatoes with salt and warm yogurt mixture.

To assemble, place mushroom mixture in prepared pie plate. Spoon potato mixture over mushroom mixture and smooth to edges, making an even layer. Make a cross-hatch design using the tines of a fork. Bake 30 to 35 minutes, or until top is lightly browned.

Makes 6 servings.

Onion-Buckwheat Mushroom Crepes

Rich-tasting with a creamy, velvet texture. The filling and topping complements the delicately flavored crepes.

CREPES

2 teaspoons olive oil
1 cup chopped onion
3/4 cup low-fat milk
2 eggs
1/2 cup buckwheat flour
1/2 cup unbleached all-purpose
  flour
1/4 teaspoon salt
Butter or soy margarine, melted,
  for cooking

FILLING

2 tablespoons butter or soy
  margarine
1/2 cup finely chopped onion
1/4 cup minced shallots
1 pound mushrooms, stems
  trimmed, thinly sliced
2 tablespoons whole-wheat flour
1/2 cup vegetable broth
1/2 cup fat-free sour cream
1 tablespoon minced fresh parsley
2/3 cup low-fat milk

Prepare crepes: Heat the oil in a medium-size, nonstick skillet over medium heat. Add onion and cook, stirring occasionally, until the onion is very tender, about 4 to 5 minutes. Place onion, milk, eggs, flours and salt in a blender container and process until very smooth. Cover and refrigerate 30 minutes.

Blend crepe batter until evenly mixed. Heat a medium nonstick skillet over medium heat until drops of water dance on the surface. Brush lightly with butter. Add 1/4 cup batter to the center of the pan and swirl pan so batter coats the bottom evenly.

Cook until bottom of crepe is browned, the top appears dry and the edges lift easily from the pan. Slide crepe onto a wire rack and repeat with remaining batter. Blend batter after making half the crepes, and add a little extra milk if the batter seems too thick. Makes 8 (7-inch) crepes.

Prepare filling: Melt the butter in a large, nonstick skillet over medium heat. Add the onion and shallots and cook, stirring frequently, until the onion is translucent, about 5 minutes. Add mushrooms and cook, stirring occasionally, until mushrooms are tender and liquid has evaporated.

Sprinkle mushrooms with flour and cook, stirring, 1 minute. Stir in broth and cook, stirring, until mixture comes to a boil and is thickened. Add sour cream and parsley, stir and remove from heat. Cover and keep warm.

Preheat oven to 375F (190C). Lightly oil a 13 x 9-inch baking dish.

To assemble crepes: Place about 3 tablespoons of filling close to one side of a crepe. Roll up crepe loosely and place in prepared baking dish. Repeat with remaining crepes.

Add milk to remaining mushroom filling. Cook, stirring, over medium heat until mixture is combined and thickened slightly. Spoon sauce over rolled crepes.

Cover baking dish with foil. Bake 10 minutes, or until heated through. Serve at once, as this dish does not hold well.

Makes 4 servings.

Root Vegetable Cobbler

Sweet potato, onion, carrots, potatoes, turnips and garlic make a savory, richly flavored combination. Topped with a cobbler crust, this is an autumn or winter main dish with substance.

2 tablespoons olive oil
1 cup chopped onion
3 cups sliced carrots
2 cups cubed russet potatoes
2 cups cubed sweet potatoes
1 cup finely diced turnips
2 garlic cloves, minced
2 tablespoons whole-wheat flour
1/2 teaspoon ground coriander
1/2 teaspoon salt
1/8 teaspoon freshly ground black
 pepper
2 cups vegetable broth

COBBLER CRUST
1 1/2 cups unbleached all-purpose
 flour
1/2 cup whole-wheat flour
1 teaspoon baking powder
1 teaspoon baking soda
1/4 teaspoon salt
Dash of freshly grated nutmeg
6 tablespoons butter or soy
 margarine, chilled, cut into
 pieces
2/3 to 3/4 cup low-fat milk

Preheat oven to 350F (175C).

Heat the oil in a nonstick Dutch oven over medium heat. Add the onion and cook, stirring occasionally, until the onion is tender and well browned, about 5 to 6 minutes. Add carrots, potatoes, sweet potatoes, turnips and garlic and cook, stirring frequently, about 3 to 4 minutes.

Sprinkle vegetables with flour, coriander, salt and pepper and cook, stirring, 1 to 2 minutes to brown flour slightly. Stir in broth and gently scrape up any flour that sticks to the bottom of the pan. Simmer vegetables, uncovered, 10 minutes, stirring.

Meanwhile, prepare dough for crust. Remove Dutch oven from heat. Carefully top with the dough. Bake 45 minutes, or until the top is browned and vegetables are tender.

Makes 8 servings.

Cobbler Crust

Place the flours, baking powder, soda, salt and nutmeg in a food processor and combine with several on-off bursts. Add butter and process until the mixture resembles coarse meal. (Butter can also be cut into the flour mixture in a large bowl by using a pastry blender.) Stir in milk just until the dough holds together.

On a lightly floured surface, roll out the dough in a circle large enough to fit over the vegetables in the Dutch oven. Cut the dough slightly larger than the Dutch oven lid. Cut a few slits in the dough to release steam.

Education

The turnip is a simple vegetable, a member of the desirable cruciferous team, which has gained status as scientists discover anticancer properties. A good source of complex carbohydrates, turnips have a peppery-sweet flavor that adds depth to dishes in which they are included. (Throw a small, cubed turnip in to cook with potatoes next time you crave "mashed," and you'll be pleasantly rewarded.) Small turnips are tender and sweet. Choose them over larger ones, which can be pithy.

Roll-Up Lasagne

An attractive dish for special occasions, and it's easier than ordinary lasagne. The sautéed onion and zucchini mixture add extra flavor and texture to the ricotta filling. Spread filling all the way to the ends of the lasagne noodles, so the rolls hold their shape.

8 ripple-edge lasagne noodles
    (about 1/2 pound)
2 teaspoons olive oil
1 cup chopped onion
2 cups shredded zucchini
1 tablespoon minced fresh parsley
1 tablespoon minced fresh basil or
    1/2 teaspoon dried basil

1 garlic clove, minced
Dash of freshly grated nutmeg
1 (15-oz.) container part-skim
    ricotta cheese
3 cups Herbed Tomato Sauce
    (page 149), or bottled sauce
1/4 cup freshly grated Parmesan
    cheese

Preheat oven to 350F (175C). Lightly oil a 13 x 9-inch nonstick cooking pan.

Bring a large pot of water to a boil. Cook lasagne noodles 10 minutes, or according to package directions. Drain and cool in a single layer on a lightly oiled, nonstick baking sheet.

Meanwhile, heat the oil in a medium, nonstick skillet over medium heat. Add the onion and cook, stirring occasionally, until the onion is translucent, about 2 to 3 minutes. Add zucchini, parsley, basil and garlic and cook, stirring, until the zucchini is wilted, 2 to 3 minutes. Dust with freshly grated nutmeg and let cool slightly. Stir together the ricotta and onion mixture.

To assemble casserole, pour 1/2 cup of tomato sauce in the bottom of prepared baking pan. Laying each lasagne noodle flat on a work surface, spread a scant 1/3 cup of the ricotta mixture along its length. Roll up and place, seam side down, in the casserole dish. Repeat with remaining lasagne noodles and ricotta mixture.

Pour the remaining sauce over and around the rolled lasagne. Cover tightly with foil. Bake 30 to 35 minutes, or until casserole is hot and bubbly. Sprinkle with Parmesan cheese and serve.

Makes 4 servings.

Lazy Lasagne

Using no-boil lasagne noodles and bottled pasta sauce makes this lasagne so easy, it can be whipped up in minutes, then put in the oven to bake while you catch up with yourself. Serve with a big green salad and crisp bread sticks.

1 1/2 cups (6 oz.) shredded
    mozzarella cheese
1 cup part-skim ricotta cheese
1/4 cup freshly grated Parmesan
    cheese
1 egg, beaten

3 tablespoons minced fresh parsley
1 tablespoon thinly sliced fresh basil
    leaves (optional)
2 cups tomato pasta sauce
6 no-boil lasagne noodles

Preheat oven to 350F (175C). Lightly oil a 9-inch-square baking pan.

Combine the mozzarella, ricotta, Parmesan, egg, parsley and basil, if using, in a medium bowl. Spoon 1/4 cup of the pasta sauce into the bottom of prepared pan. Top with 2 noodles and spread noodles with about half of the cheese mixture.

Spoon 3/4 cup of the sauce over top. Layer with 2 noodles going in the opposite direction from the bottom layer of noodles. Spread with remaining cheese mixture. Top with remaining noodles going in the same direction as the bottom layer. Spoon remaining sauce over noodles.

Cover pan with foil and bake 30 minutes. Remove foil and bake 15 to 20 minutes, or until noodles are tender.

Makes 4 servings.

Education

Layering lasagne noodles in opposing directions, crosshatch-style, lends stability to the final dish. You will find the lasagne holds together better for serving. Be sure the dish is big enough to accommodate the noodles, without touching each other or the sides of the casserole, as the noodles will expand.

Satisfying
Stovetop Favorites

This chapter offers a good selection of foods influenced by different cultures, dishes from the Middle East to southern Europe to the Far East to our southern neighbor, Mexico. Because many ethnic groups eat a vegetarian or almost-vegetarian diet, these dishes fit easily into a vegetarian cookbook.

Also, because many countries have limited supplies of fuel for cooking, dishes that can be prepared easily over a heat source make good economic sense.

Falafel

Middle Eastern falafel makes a tempting meatless burger, especially when embellished with tahini sauce and a tomato topping.

1/3 cup bulgur

1/3 cup boiling water

2 tablespoons fresh lemon juice

Tahini Sauce (see below)

1 cup peeled, seeded and chopped
    tomatoes

1/3 cup peeled, seeded and finely
    diced cucumber

1 tablespoon minced green bell
    pepper

1 tablespoon minced or grated
    onion

1 tablespoon minced fresh parsley

2 eggs

1 1/2 cups cooked (or 15-oz. can)
    chickpeas, drained

1/2 cup dry bread crumbs

2 tablespoons packed parsley leaves

2 tablespoons packed cilantro leaves

1 teaspoon ground cumin

1/2 teaspoon ground coriander

1/2 teaspoon red chili pepper flakes

1/2 teaspoon salt

1/8 teaspoon freshly ground black
    pepper

About 1 tablespoon olive oil

TAHINI SAUCE

3 tablespoons sesame tahini

2 tablespoons water

2 tablespoons fresh lemon juice

1 teaspoon olive oil

1 small garlic clove, minced

Dash of ground hot red pepper

Dash of salt

Place bulgur in a small bowl and add boiling water and lemon juice. Cover and set aside 20 minutes, or until the liquid is absorbed.

Meanwhile, prepare Tahini Sauce and set aside. Combine tomatoes, cucumber, bell pepper, onion and minced parsley and set aside.

Combine soaked bulgur, eggs, chickpeas, bread crumbs, parsley, cilantro, cumin, coriander, pepper flakes, salt and black pepper in a food processor and process until smooth. Shape mixture into 4 patties.

Heat oil in a large nonstick skillet over medium heat. Add patties and cook, turning to brown both sides, until patties are cooked through, about 10 minutes. To serve, drizzle with sauce and top with tomato mixture.

Makes 4 servings.

Tahini Sauce

Combine sesame tahini, water, lemon juice, olive oil, garlic, cayenne and salt in a small bowl.

Makes about 1/2 cup.

Mock Burger

This vegetarian version can be used in anything from tacos to spaghetti sauce, filled pasta shells to casseroles. The basic recipe is unseasoned, so you can add herbs and/or spice as the finished dish warrants. The brown lentils provide more of the texture associated with ground meat.

2 small carrots
1 medium onion
1 tablespoon olive oil
3/4 cup brown lentils, picked
  over and rinsed

1/4 cup red lentils (see Note,
  page 124), picked over and
  rinsed
2 cups water

Trim carrots and cut into large chunks. Place in a food processor and process until coarsely grated. Peel and quarter onion. Add to carrots and process until finely grated.

Heat the oil in a large skillet over medium heat. Add the carrot mixture and cook, stirring frequently, until mixture is soft and lightly browned, about 5 minutes.

Add lentils and water and bring to a boil. Reduce heat, cover and simmer about 20 minutes, or just until the lentils are firm-tender and the water is absorbed. Do not overcook or lentils will become mushy.

Makes 3 cups, about 12 servings, depending upon recipes.

Lentil & Bean Tacos Supreme

Make the filling in your skillet, and, as it cooks, prepare toppings. Crisp the taco shells in the oven, then serve buffet style.

2 tablespoons olive oil
1 onion, chopped
1 jalapeño chile, seeded and minced
2 garlic cloves, minced
1 tablespoon chili powder
2 teaspoons ground coriander
1 teaspoon ground cumin
1/2 teaspoon salt
1/2 cup brown lentils, picked over
    and rinsed
1 tablespoon bulgur
2 1/2 cups vegetable broth
1 2/3 cups cooked (or 15-oz. can)
    red kidney beans, drained

2 tablespoons tomato paste
2 tablespoons minced fresh parsley
1 tablespoon minced fresh cilantro
    leaves
1 teaspoon molasses
12 taco shells
1 cup Garlic Lovers' Salsa Cruda
    (see page 3) or bottled salsa
1/2 cup low-fat yogurt (optional)
2 cups thinly sliced romaine or
    iceberg lettuce leaves
1 1/2 cups (6 oz.) shredded
    Cheddar cheese

Heat oil in a medium, nonstick skillet over medium heat. Add the onion, chile and garlic and cook, stirring occasionally, until the onion is translucent, about 3 to 4 minutes. Add chili powder, coriander, cumin and salt and cook, stirring, 1 minute.

Stir in lentils, bulgur and broth, and bring to a boil. Reduce heat, partially cover and simmer 30 minutes, or until the lentils are tender and the liquid is mostly absorbed.

Preheat oven to 350F (175C). Lightly mash kidney beans in a bowl and add to skillet. Stir in tomato paste, parsley, cilantro and molasses and cook 2 to 3 minutes, until flavors are blended.

Place taco shells on a baking sheet and warm in the oven about 5 minutes. To serve, divide bean mixture among taco shells and top with salsa, yogurt, if desired, lettuce and cheese.

Makes 6 servings.

Middle Eastern Chickpeas

A flavorful combination of spinach and tomato can best be served with a side dish of orzo pasta.

2 (10-oz.) bags fresh spinach
2 tablespoons olive oil
1 large onion, chopped
3 garlic cloves, minced
3 1/2 cups cooked [or 2 (15-oz.) cans] chickpeas

1 cup peeled, seeded and diced tomatoes
1/4 cup tomato paste
1/4 teaspoon salt
1/4 teaspoon freshly ground black pepper

Remove stems and rinse spinach. Cut spinach leaves into thin strips. Set aside.

Heat the oil in a large, nonstick skillet over medium heat. Add the onion and garlic, and cook, stirring occasionally, until onion is translucent, about 3 to 4 minutes. Drain liquid from chickpeas and add 1 cup of drained liquid to the onion mixture. Stir in tomatoes, tomato paste, salt and pepper and bring to a boil.

Stir in spinach until wilted. Stir in drained chickpeas. Simmer, covered, 25 minutes, or until vegetables are tender.

Makes 6 servings.

Education

Eating tomatoes or other vitamin C–rich foods with chickpeas and other legumes is an iron-clad technique for making the most of them nutritionally. For while beans are a good source of iron, it is not absorbed as well as iron from meat and other animal foods. Adding foods rich in vitamin C increases your ability to absorb the iron, helping you make the most of your meal.

Vegetable Rice with Cannellini

White beans and vegetable-enhanced rice make a main dish with ease. Serve with garlic bread and a green salad.

2 tablespoons olive oil
1 large onion, chopped
1 yellow bell pepper, diced
1 cup converted rice
2 cups water
1 (14-oz.) can Italian-style stewed
    tomatoes
1 3/4 cups cooked (or 15-oz. can)
    cannellini beans, drained
    (white kidney beans)
1 cup frozen whole-kernel corn

1 tablespoon minced fresh basil or
    1/2 teaspoon dried basil,
    crumbled
1 teaspoon fresh marjoram leaves or
    1/2 teaspoon dried marjoram,
    crumbled
1/2 teaspoon fresh thyme leaves or
    1/8 teaspoon dried thyme
1/4 teaspoon salt
1/4 teaspoon freshly ground black
    pepper

Heat the oil in a large, nonstick skillet over medium heat. Add the onion and bell pepper and cook, stirring, until the onion is translucent, about 3 to 4 minutes. Stir in rice, add water and bring to a boil. Reduce heat, cover and simmer 20 minutes, or until rice is nearly tender.

Pour stewed tomatoes into a blender and process until pureed. Add to the rice along with beans, corn, basil, marjoram, thyme, salt and pepper, and bring to a boil. Reduce heat, cover and simmer 10 minutes. Serve hot.

Makes 4 servings.

Education

Much of the world's population survives on beans (and bean products, such as tofu) and rice. The legumes and grain complement each other to make a top-quality protein that eliminates the need for meat, which is scarce in many parts of the world.

Broccoli Stir-Fry

Broccoli complements the flavor of cashews. Prepare vegetables while the rice cooks.

1 1/2 cups converted rice
3 cups water
2 large stalks broccoli
2 tablespoons olive oil
1 large onion, chopped
1 small carrot, thinly sliced on
    diagonal

2 garlic cloves, minced
1/2 yellow or red bell pepper, diced
2 tablespoons teriyaki sauce
2 tablespoons fresh orange juice
Dash of ground cayenne
1/3 cup chopped cashews

Bring rice and water to a boil in a large saucepan. Reduce heat, cover and simmer 25 minutes, or until water is absorbed. Keep covered and set aside.

Meanwhile, peel broccoli stems and cut into thin, 2-inch sticks. Chop broccoli tops.

Heat the oil in a large, nonstick skillet over medium heat. Add the onion and cook, stirring frequently, until onion is translucent, about 3 to 4 minutes. Add carrot and garlic and cook, stirring, 2 to 3 minutes, or just until carrots are crisp-tender. Add bell pepper and broccoli, teriyaki sauce, orange juice and cayenne and stir to combine. Cover and simmer 4 to 5 minutes, or just until vegetables are firm-tender. Stir in cashews. Serve over rice.

Makes 4 servings.

Education

Choose fresh broccoli if excess sodium in your diet is a concern. Frozen broccoli contains twice as much sodium as fresh. To make sure fresh broccoli tastes best, choose slender stalks with tightly closed buds that are green. Yellow tops reveal that the broccoli is past its prime.

Tex-Mex Taco Dinner

Serve with steamed broccoli and corn. Don't be alarmed by the number of ingredients—the filling is actually simple—most of the ingredients are seasonings.

1 tablespoon olive oil
1 large onion, chopped
5 garlic cloves, minced
1 teaspoon chili powder
1/2 teaspoon *each* salt, ground
    cumin and ground coriander
1/4 teaspoon *each* dried thyme
    and dried oregano
2 small carrots, coarsely chopped
1 celery stalk, coarsely chopped
1/2 cup red lentils (see Note,
    page 124), picked over and rinsed

1 1/2 cups water
8 (8-inch) flour tortillas
1 2/3 cups cooked (or 15-oz. can)
    kidney beans, drained
1 (8-oz.) can tomato sauce
1 tomato, peeled, seeded and
    chopped
3 green onions, thinly sliced on
    diagonal
1/2 cup taco sauce, warmed
2 cups thinly sliced romaine lettuce

Heat the oil in a large, nonstick skillet over medium heat. Add the onion and cook, stirring frequently, until onion is tender and lightly browned, about 5 to 7 minutes. Add garlic and cook, stirring, 1 minute. Add chili powder, salt, cumin, coriander, thyme and oregano and stir until thoroughly combined. Remove from heat.

Process carrots and celery in a food processor until finely chopped. Add carrot mixture, lentils and water to onion mixture and bring to a boil over high heat. Reduce heat and simmer, uncovered, until lentils are tender and liquid is evaporated, about 10 minutes.

Place the tortillas on a microwave-safe plate. Top with a damp paper towel and cover with plastic wrap, wrapping around plate. Microwave on HIGH 1 1/2 minutes. Keep wrapped until ready to serve. To heat tortillas in the oven, brush each lightly with oil, stack and wrap in foil. Bake 15 minutes at 300F (150C). Keep wrapped until ready to use.

Stir kidney beans and tomato sauce into lentil mixture and simmer until heated through.

To serve, divide bean mixture among flour tortillas. Sprinkle each with chopped tomato and green onions. Fold or roll up. Spoon taco sauce over tacos and top with lettuce.

Makes 4 servings.

Variations

This bean filling is so versatile, it can be used in a variety of ways. Hollow out zucchini halves and steam until tender. Fill with bean mixture, top with tomato sauce and microwave or bake until hot. Mix the bean filling with cooked rice and use it to stuff steamed green bell peppers, too. Bake as with zucchini.

Moroccan Curried Squash with Golden Raisins

Serve over brown rice, couscous or bulgur for a satisfying, flavorful meal.

1 tablespoon canola oil	2 cups water
1 cup sliced onion	2 cups peeled, diced acorn or
2 garlic cloves, minced	butternut squash or sweet
1 teaspoon ground cinnamon	potato
1 teaspoon ground coriander	1 2/3 cups cooked (or 15-oz. can)
1/2 teaspoon ground ginger	chickpeas, drained
1/4 teaspoon ground turmeric	1/2 cup golden raisins

Heat the oil in a large, nonstick skillet over medium heat. Add the onion and cook, stirring frequently, until onion is crisp-tender, about 6 minutes. Add garlic, cinnamon, coriander, ginger and turmeric and cook, stirring, 1 minute.

Stir in remaining ingredients and bring to a boil. Cover and simmer until squash is tender, about 10 to 15 minutes. Serve over cooked rice.

Makes 4 servings.

Education

Winter squash, such as butternut, is so rich in beta carotene that one serving, roughly 3 1/2 ounces, can provide 100 percent of the recommended daily allowance. And that's beta carotene with convenience, since winter squash keeps so well. Unblemished squash can be stored three months or more in a cool, dry place.

Egyptian Lentil Couscous with Raisins & Apricots

Delicately flavored and colorful, this main dish is a meal in itself. Serve with optional yogurt sauce (see Variation) on the greens, if desired.

1 tablespoon canola oil
1 large onion, halved and thinly
    sliced crosswise
1 tablespoon minced garlic
1 teaspoon ground coriander
1/2 teaspoon ground cardamom
1 cup brown lentils, picked over
    and rinsed
4 cups water
1/3 cup raisins

1/4 cup finely chopped dried
    apricots
1 cup shredded carrots
3/4 cup uncooked couscous
1/4 teaspoon salt
1 cup packed thinly sliced spinach,
    for garnish
1 tablespoon minced fresh cilantro
    leaves, for garnish

Heat the oil in a large, nonstick skillet over medium heat. Add the onion and cook, stirring frequently, until onion is translucent, about 3 to 4 minutes. Add garlic, coriander and cardamom and cook, stirring, 1 minute. Add lentils, water, raisins and apricots and bring to a boil. Reduce heat, cover and simmer 20 to 25 minutes or until the lentils are just firm-tender.

Stir in carrots, couscous and salt. Return to a boil and cook 5 to 6 minutes, stirring once or twice, or until liquid is absorbed and carrots are firm-tender.

To serve, mound lentil couscous on warmed individual plates. Toss together spinach and cilantro. Arrange a small ring of the greens around each mound of lentils.

Makes 4 servings.

Variation

A yogurt sauce can be drizzled over the spinach mixture. For yogurt sauce, combine 2/3 cup plain low-fat yogurt with 1 minced garlic clove, 1/8 teaspoon crumbled dried mint leaves and a dash of salt.

Education

Don't think of just salads when you're serving greens. Deep emerald leaves of spinach or watercress can be a gorgeous (and edible) garnish for main dishes, as well. The creative touch adds more than just good looks; it also provides chlorophyll, beta carotene and vitamin C.

Midwest Lentils with Chickpeas

A medley of sizes, shapes and textures. Serve with quick brown rice or couscous for a convenient dinner.

2 tablespoons olive oil
1 1/2 cups diced carrots
1 cup diced celery
1/2 cup sliced green onions
1 garlic clove, minced
1/2 teaspoon ground coriander
1 cup brown lentils, picked over
  and rinsed

2 1/2 cups vegetable broth
1 bay leaf
1 2/3 cups cooked (or 15-oz. can)
  chickpeas, drained
2 cups cherry tomatoes, halved
1/4 cup minced fresh parsley
2 tablespoons minced fresh
  cilantro (optional)

Heat the oil in a nonstick Dutch oven over medium heat. Add the carrots, celery and green onions and cook, stirring occasionally, until the vegetables are firm-tender, about 4 to 5 minutes. Stir in garlic and coriander and cook, stirring, 1 minute.

Add lentils, broth and bay leaf and bring to a boil. Reduce heat, cover and simmer 20 to 25 minutes, or until lentils are nearly tender.

Remove bay leaf. Add chickpeas, cherry tomatoes, parsley and cilantro, if desired. Cook 5 to 8 minutes, uncovered, until flavors are blended.

Makes 4 servings.

Mediterranean Ratatouille with Double-Corn Polenta

Put a lid on it . . . that's the easiest way to make ratatouille. After briefly sautéing the onions, stir in remaining ingredients and simmer for a perfect end-of-summer main dish!

1 tablespoon olive oil
2 onions, halved and thinly sliced
    crosswise
4 garlic cloves, thinly sliced
1 small eggplant, cut into
    1/4-inch-thick slices (see
    Note, opposite), then cut into
    1/2-inch-wide strips
1 medium zucchini, cut into
    2 x 1/2-inch strips
1 yellow or orange bell pepper, cut
    into 2 x 1/2-inch strips
3 tomatoes, peeled, seeded and
    chopped
2/3 cup eight-vegetable juice
2 tablespoons minced fresh parsley

1 teaspoon minced fresh basil or
    1/2 teaspoon dried basil,
    crumbled
1 teaspoon minced fresh oregano
    or 1/4 teaspoon dried oregano,
    crumbled
1/4 teaspoon salt
1/8 teaspoon freshly ground black
    pepper
2/3 cup stone-ground yellow
    cornmeal
3 cups water
1 tablespoon soy margarine or butter
3 ears cooked corn, kernels
    removed, or 1 cup frozen white
    shoepeg corn, thawed

Heat the oil in a nonstick Dutch oven over medium-high heat. Add the onions and cook, stirring occasionally, until the onions just begin to turn golden, about 5 to 7 minutes. Add garlic and cook, stirring, 1 minute.

Add eggplant, zucchini, bell pepper, tomatoes, juice, parsley, basil, oregano, salt and black pepper. Stir until combined and bring to a boil. Reduce heat, cover and simmer 30 minutes, stirring once, or until the vegetables are tender.

Meanwhile, combine the cornmeal, water and margarine or butter in a large microwave-safe bowl. Microwave, uncovered, on HIGH 3 minutes or until thickened. Stir well. Microwave, uncovered, on HIGH 3 minutes. Stir in corn and microwave, uncovered, on HIGH 2 minutes.

To serve, divide polenta among 6 large, flat soup bowls, or serve on deep plates. Top with the ratatouille.

Makes 6 servings.

Variation

Ratatouille can also be served over couscous or angel hair pasta.

Note Many cooks like to remove from eggplant the excess liquid, which can be bitter, before cooking. To do this, salt slices on both sides. Arrange in 4 equal stacks on a cutting board slanted to drain into the sink. Weight down the slices with a Dutch oven filled with water. Let stand anywhere from 20 minutes to 2 hours. Rinse off salt well before using.

Education

Ratatouille can tip the scale with fat if the eggplant is fried separately before adding it to the vegetable mixture. Eggplant can absorb prodigious amounts of oil—four times the amount that potatoes can, which is already substantial.

Thin Spaghetti with Red Lentil Sauce

Small red beans and red lentils provide a hearty boost of flavor (and protein) to your favorite pasta sauce.

1/3 cup red lentils (see Note, below), picked over and rinsed
1 garlic clove, sliced
2/3 cup water
1 pound thin spaghetti
1 2/3 cups cooked (or 15-oz. can) small red beans, drained

3 1/2 cups Herbed Tomato Sauce (page 149) or bottled sauce
1/2 teaspoon dried basil
2 tablespoons minced fresh parsley, for garnish

Bring lentils, garlic and water to a boil in a small saucepan over medium-high heat. Reduce heat and simmer 8 to 10 minutes, uncovered, or until lentils are firm-tender.

Meanwhile, bring a large pot of water to a boil. Add pasta and stir. Return to a boil, then reduce heat and cook 7 to 10 minutes, or until firm-tender.

Place beans, pasta sauce, basil and lentils with any remaining liquid in a large saucepan. Bring to a boil over medium-high heat. Stir, reduce heat, cover and simmer 10 minutes.

Place pasta on a large serving plate, or individual plates, and top with sauce. Garnish with minced fresh parsley.

Makes 6 servings.

Note If your supermarket does not have red lentils, regular brown lentils can be substituted. Cook an additional 5 minutes, adding extra water, if needed.

Education

For emotional equilibrium, balance your diet with lentils. Lentils are rich in B vitamins, so important to nerve function.

Red Lentil Tostadas

Topped with diced tomatoes, cucumbers, green onions and sprouts, this is a main-dish-and-salad-in-one.

3/4 cup red lentils (see Note, page 124), picked over and rinsed
1 tablespoon olive oil
1 onion, finely chopped
1 carrot, finely chopped
2 garlic cloves, minced
1 teaspoon chili powder
1/2 teaspoon ground cumin
1/4 teaspoon ground coriander
1/4 teaspoon salt

3/4 cup eight-vegetable juice
3/4 cup vegetable broth
1 tablespoon minced fresh cilantro
8 tostada shells
1 large tomato, peeled, seeded and diced
1 small cucumber, peeled, seeded and diced
2 green onions, thinly sliced
1 cup alfalfa sprouts

Place the lentils in a medium bowl and cover with boiling water. Set aside.

Heat oil in a large, nonstick saucepan over medium heat. Add onion and carrot and cook, stirring frequently, until onion begins to turn golden, about 5 to 7 minutes. Add garlic, chili powder, cumin, coriander and salt and cook, stirring, 1 minute. Add lentils, juice and broth and bring mixture to a boil. Reduce heat and simmer, uncovered, until lentils are tender and liquid is absorbed, about 10 to 12 minutes. Remove from heat and stir in cilantro.

Warm tostada shells according to package directions. To serve, bring tostadas to the table in a basket, wrapped in a cloth napkin. Transfer lentil mixture to a serving bowl and place tomato, cucumber, onions and sprouts in small dishes. Let each person fill a tostada with some of the lentil mixture on a tostada and top with choice of vegetables.

Makes 4 servings.

Education

Crisp salad ingredients can add a contrast of texture and flavor to a variety of bean dishes. Toss some thinly sliced greens, diced tomatoes and/or green onions on a variety of your favorite hot meals. It's even quicker than a side salad, and there's no need for potentially high-fat dressings.

Linguine with Oyster Mushroom Sauce

You can use regular button mushrooms, or another variety, depending on your spirit of adventure. The oyster mushrooms, however, yield a very special flavor.

2 cups (about 1/4 lb.) oyster
    mushrooms
1 tablespoon olive oil
1 garlic clove, minced
1/2 teaspoon salt
Dash of freshly grated nutmeg
1/2 cup vegetable broth

1/2 cup tomato sauce
1/2 cup low-fat milk
2 tablespoons minced fresh parsley
3/4 pound linguine
1/4 cup freshly grated Parmesan
    cheese (optional)

Dice mushrooms. Heat the oil in a large, nonstick skillet over medium-high heat. Add mushrooms and cook, stirring occasionally, 4 to 5 minutes. Add garlic, salt and nutmeg and cook, stirring, 1 minute.

Add broth, tomato sauce and milk and bring to a boil. Reduce heat, cover and simmer 10 minutes, or until the mushrooms are tender. Stir in parsley and remove from heat.

While mushrooms are cooking, bring a large pot of water to a boil. Cook linguine until firm-tender, about 9 to 11 minutes. Drain.

Place linguine in a warmed serving bowl, and top with mushroom sauce. Add grated Parmesan, if desired.

Makes 4 servings.

Education

Oyster mushrooms have a flavor, color and texture reminiscent of seafood. Diced, these mushrooms yield a sauce that resembles clam sauce in appearance. Sometimes those who are new to vegetarian cooking enjoy eating dishes that look familiar. The linguini, above, may put newcomers at ease. The delicious flavor, however, is bound to win them over.

Three-Bean Skillet Chili

Set out a selection of toppings so each person can choose his or her favorites. A creative choice of toppings enhances the appearance and flavor of any chili.

2 tablespoons olive oil
1 large onion, chopped
1 green bell pepper, chopped
2 large garlic cloves, minced
2 tablespoons chili powder
1 teaspoon ground cumin
1 teaspoon ground coriander
1/2 teaspoon dried oregano
1/2 teaspoon salt
3 cups vegetable broth or water
1 cup brown lentils, picked over
    and rinsed
1 (28-oz.) can chopped tomatoes
    with juice
1 (8-oz.) can tomato sauce
1 2/3 cup (or 15-oz. can) red kidney
    beans, drained

1 2/3 cup (or 15-oz. can) pink
    beans, drained
1/4 cup minced fresh parsley
2 tablespoons minced fresh cilantro

SUGGESTED TOPPINGS
Shredded sharp Cheddar cheese
Shredded lettuce
Shredded zucchini
Chopped green olives
Chopped black olives
Fat-free sour cream
Salsa
Peeled, seeded and diced tomatoes
Diced avocado
Alfalfa sprouts

Heat the oil in a large skillet over medium heat. Add the onion and bell pepper and cook, stirring occasionally, until the onion is translucent, about 4 to 5 minutes. Add garlic, chili powder, cumin, coriander, oregano and salt and cook, stirring, 1 minute.

Add broth and lentils and bring to a boil. Reduce heat, cover and simmer 20 to 25 minutes, or until lentils are firm-tender.

Add tomatoes and tomato sauce and return to a boil. Simmer, uncovered, 15 minutes or until slightly thickened. Stir in beans, parsley and cilantro and heat through. Serve with choice of toppings on the side.

Makes 8 servings.

Taj Lentils with Rice & Chutney

An easy lentil curry, accompanied with spicy chutney, makes a satisfying meal with quick brown rice. For a cooling accompaniment, stir peeled, seeded and shredded cucumber into yogurt.

1 tablespoon canola oil
2 onions, halved and sliced
2 teaspoons minced fresh garlic
2 teaspoons curry powder
1/2 teaspoon ground fennel seeds
1/2 teaspoon ground coriander
1/2 teaspoon salt
1/2 teaspoon freshly ground black pepper
1/2 teaspoon mustard seeds
3 1/4 cups vegetable broth

1 1/2 cups brown lentils, picked over and rinsed
1/2 cup golden raisins
2 1/4 cups water
2 cups instant brown rice
1/3 cup minced fresh parsley
1 tablespoon minced fresh cilantro or mint (optional)
Chutney, for garnish
Chopped peanuts, for garnish

Heat the oil in a nonstick Dutch oven over medium heat. Add the onions and cook, stirring frequently, until the onions are translucent, about 4 to 5 minutes. Add garlic, curry, fennel, coriander, salt, pepper and mustard seeds, and cook, stirring, 1 minute.

Add broth, lentils and raisins and bring to a boil. Reduce heat, cover and simmer 20 to 25 minutes, or until lentils are tender.

Meanwhile, bring water to a boil and stir in rice. Reduce heat, cover and simmer 10 minutes, or until water is absorbed. Cover and set aside.

Stir parsley and cilantro or mint, if using, into cooked lentils, and cook 1 to 2 minutes, or just until herbs are wilted. Serve over rice and provide side dishes of chutney and peanuts.

Makes 4 servings.

Education

While peanuts are high in complementary protein and can be an important part of a vegetarian diet, they are also high in fat and should be used sparingly. Here a small amount of chopped peanuts can be used as a garnish, providing flavor and valuable protein without a lot of fat.

Eggplant Curry

Serve over rice, as they would in India, or cross culture by serving with couscous or orzo pasta. The creamy curry will complement any of these.

2 tablespoons canola oil
1 large onion, halved lengthwise
    and thinly sliced crosswise
1 teaspoon minced garlic
1 teaspoon *each* ground coriander
    and ground cumin
1/2 teaspoon chili powder
1/4 teaspoon *each* mustard seeds,
    ground turmeric and ground
    cinnamon

1 (1 1/4- to 1 3/4-lb.) eggplant,
    peeled and diced
1/2 red bell pepper, cut into thin
    strips
1/2 teaspoon salt
1 cup low-fat milk or vegetable
    broth
2 tablespoons minced fresh parsley
2 tablespoons minced fresh cilantro
    leaves

Heat the oil in a nonstick Dutch oven over medium-high heat. Add the onion and cook, stirring frequently, until the onion is translucent, about 3 to 4 minutes. Add the garlic, coriander, cumin, chili powder, mustard seeds, turmeric and cinnamon and cook, stirring, 1 minute.

Stir in eggplant, bell pepper and salt and cook, stirring frequently, 5 minutes or until the eggplant is firm-tender. Stir in milk and bring to a boil. Reduce heat, cover and simmer 15 to 20 minutes until the vegetables are tender. Stir in parsley and cilantro and continue to cook 1 to 2 minutes.

Makes 4 servings.

Indonesian Fried Rice

Nasi goreng *or Indonesian fried rice features a variety of vegetables and the additional protein of egg. It is traditionally served in Holland with sliced gherkins and tomato wedges.*

1 1/3 cups long-grain converted rice
2 eggs
3 tablespoons cold water
3 tablespoons canola oil
2 cups thinly sliced leeks
2 onions, finely chopped
1/2 jalapeño chile, seeded and
    minced

2 garlic cloves, minced
1/2 teaspoon ground coriander
1/2 teaspoon ground cumin
3 tablespoons soy sauce
Gherkins, for garnish
Tomato wedges, for garnish

Bring a large saucepan half-filled with water to a boil. Add rice and cook 18 minutes. Drain and rinse under cold running water.

Beat the eggs well with the water. Heat about 1 tablespoon of the oil in a large, nonstick skillet over medium heat. Add eggs, swirling the pan slightly so that the eggs cook in one thin layer. Do not stir. When the eggs are just set, remove the pan from the heat and set aside to finish cooking with the heat from the pan.

Heat remaining oil in a nonstick Dutch oven over medium heat. Add the leeks, onions and chile and cook, stirring frequently, until the mixture begins to turn golden, about 12 to 15 minutes. Add garlic, coriander and cumin and cook, stirring, 1 minute. Add rice and soy sauce and cook, stirring frequently, 5 minutes, until rice is heated through and flavors are blended.

Cut cooked egg into 2-inch strips and stir into the fried rice. Garnish with gherkins and tomato wedges.

Makes 4 servings.

Variation

If you don't want to add eggs, you can increase the protein content of *nasi goreng* by serving it with Peanut Sauce (opposite).

Education

Leeks are a popular vegetable in Holland and add a special flavor to Indonesian-inspired fried rice, which is often served there as a light supper. Of particular interest to vegetarians is the fact that leeks are a good source of iron.

Peanut Sauce

Peanut sauce can be served with Indonesian Fried Rice (opposite) or other vegetarian main dishes. Made with complementary nuts and milk, it is protein rich.

1/2 cup creamy peanut butter
10 tablespoons low-fat milk

1/4 cup tomato sauce
Dash of salt

Place the peanut butter in a small bowl. Add the milk 1 tablespoon at a time, stirring well after each addition. Stir in tomato sauce and salt. Warm in microwave about 1 minute on HIGH before serving.

Makes about 1 1/2 cups.

Lentil-Potato Curry with Lemon Rice

Spicy red lentils with potato tops a colorful lemon rice for an exotic main dish. Yellow tomatoes, if you have them, highlight the golden colors of this dish.

3 cups vegetable broth
1 cup split red lentils (see Note, page 124), picked over and rinsed
1 cup finely chopped onion
1 (1 1/2-inch) piece gingerroot, peeled and thickly sliced
2 large garlic cloves, peeled
1/2 jalapeño chile, seeded
2 tablespoons canola oil
1 teaspoon ground coriander
1/2 teaspoon ground cumin
1/2 teaspoon ground fennel seeds
1/4 teaspoon chili powder

2 medium potatoes, peeled and diced
2 large yellow or red tomatoes, peeled, seeded and chopped
1 teaspoon salt
2 tablespoons minced fresh parsley
2 tablespoons minced fresh cilantro leaves
1 1/2 cups long-grain converted rice
3 1/4 cups water
2 tablespoons fresh lemon juice
1 tablespoon soy margarine or butter
1/2 teaspoon grated lemon peel
1/4 teaspoon ground turmeric

Heat broth in a large saucepan over medium-high heat. Add lentils and onion and bring to a boil. Reduce heat, cover and simmer 30 minutes.

Combine the gingerroot, garlic and chile in a food processor and process until finely chopped. Heat the oil in a small, nonstick skillet over medium heat. Add gingerroot mixture and cook, stirring constantly, 1 minute. Add the coriander, cumin, fennel and chili powder and cook, stirring, 1 minute.

Add the gingerroot mixture, potatoes, tomatoes and salt to the lentils. Stir and return to a boil. Reduce heat, cover and simmer 25 minutes, stirring occasionally, until potatoes are tender. Stir in parsley and cilantro, cover and set aside.

Meanwhile, combine the rice, water, lemon juice, margarine, lemon peel and turmeric in a large saucepan and bring to a boil. Reduce heat, cover and simmer 20 to 25 minutes, until rice is tender and water is absorbed. Serve curry over rice.

Makes 6 servings.

Variation

Substitute 1 (16-oz.) can chopped tomatoes with juice for the fresh tomatoes.

Mushroom & Green Pepper Goulash

Creamy and luscious, but not high in fat, this smooth, saucy goulash is a winner!

1 tablespoon butter or soy
    margarine
1 tablespoon olive oil
4 onions, halved and thinly sliced
    crosswise
1 green bell pepper, peeled and
    diced
2 pounds button mushrooms,
    quartered
2 garlic cloves, minced
1 (6-oz.) can tomato paste

2 cups vegetable broth
1 cup peeled, seeded and chopped
    tomatoes
1 tablespoon balsamic vinegar
2 teaspoons paprika
1/2 teaspoon salt
1/4 teaspoon freshly ground black
    pepper
1 pound egg noodles
1/2 cup fat-free sour cream
2 tablespoons minced fresh parsley

Heat the butter and oil in a nonstick Dutch oven over medium heat. Add the onions and bell pepper and cook, stirring occasionally, until the onion begins to brown, about 7 to 8 minutes. Add mushrooms and garlic and cook until mushrooms release moisture, and are slightly tender, about 10 to 15 minutes.

Stir in tomato paste until well combined. Add broth, tomatoes, vinegar, paprika, salt and black pepper and bring to a boil. Reduce heat, cover and simmer 30 minutes.

Bring a large pot of water to a boil. Add noodles about 10 minutes before goulash is done cooking. Return to a boil. Cook, uncovered, 7 to 8 minutes, or until firm-tender.

Remove goulash from heat. Stir in sour cream and parsley. Serve over hot noodles.

Makes 6 servings.

Jalapeño Cornmeal Pancakes with Fresh Cilantro Salsa

Serve with green beans for a rollicking plate full of color. If you like, add a dollop of fat-free sour cream to top pancakes, then spoon on salsa. Olé!

Salsa (see opposite)
About 3 teaspoons olive oil
1/4 cup chopped onion
1/4 cup diced red bell pepper
1 garlic clove, halved
1/2 jalapeño chile, seeded
1 cup fresh or frozen corn kernels
2 eggs, beaten
1 cup stone-ground cornmeal
1 cup unbleached all-purpose flour
1 teaspoon baking powder
1/2 teaspoon baking soda
1/2 teaspoon salt
1/8 teaspoon ground coriander
1 1/4 to 1 3/4 cups buttermilk

FRESH CILANTRO SALSA
2 cups peeled, seeded and diced
   tomatoes
1/4 cup finely chopped red onion
1/4 cup minced fresh cilantro leaves
1 tablespoon fresh lime juice

Prepare salsa and set aside.

Heat 1 teaspoon of the oil in a small, nonstick skillet over medium heat. Add the onion, bell pepper, garlic and chile and cook, stirring occasionally, until the onion is translucent, about 1 to 2 minutes. Stir in corn and heat through. Let cool slightly.

Place the corn mixture in a food processor. Add eggs and process until vegetables are finely chopped. Add cornmeal, flour, baking powder and soda, salt, coriander, and 1 1/4 cups of the buttermilk. Process until smooth, scraping down the sides of the bowl as necessary. Add additional buttermilk, as necessary, to make batter the consistency of heavy cream.

Heat remaining oil in a griddle or large skillet over medium heat. Spoon batter into 4-inch pancakes. When bottoms are set and browned, turn to brown remaining side. Cook until cooked through and browned, about 10 minutes.

Serve with salsa.

Makes 4 servings.

Fresh Cilantro Salsa

Combine tomatoes, onion, cilantro and lime juice in a serving bowl.

Makes abut 2 1/2 cups.

Vegetable Frittata

This dish is a terrific way to use up small amounts of cooked vegetables. Just combine enough veggies to equal 1 to 1 1/2 cups and heat in the skillet before adding eggs. Otherwise, briefly cook fresh or frozen vegetables, as below, then sauté.

1 cup diced potatoes
1/2 cup diced celery
1 tablespoon diced onion
1 tablespoon diced red bell pepper
2 teaspoons olive oil
1 tablespoon minced fresh parsley,
    plus extra for garnish

3 eggs, beaten
1/4 cup (1 oz.) shredded Cheddar
    cheese
Thin tomato slices, for garnish

Combine the potatoes, celery, onion and bell pepper in a medium saucepan over medium heat. Add enough water to cover and bring to a boil. Reduce heat, cover and simmer 10 to 15 minutes, or until the vegetables are firm-tender. Drain.

Preheat broiler. Heat the oil in a medium nonstick skillet over medium heat. Add the vegetables and cook, stirring frequently, until they begin to brown, about 3 to 4 minutes. Stir in the parsley, and spread the vegetables out in an even layer across the bottom of the pan.

Reduce heat to low. Pour the eggs over the vegetables and allow to set without stirring. When the bottom of the eggs are cooked and only the top is runny, sprinkle with the cheese.

Place under hot broiler until cheese melts and the frittata is golden. To serve, cut in half. Garnish with tomato slices and sprinkle with parsley.

Makes 2 servings.

Education

Garnish for health: Adding sliced or cubed vegetables and fruits to enhance a meal gives family and guests more variety and nutrition. Parsley, for example, is one of the richest vegetable sources of vitamin A. By adding a sprinkling of fresh parsley to finished dishes, you will provide more of this important immune-enhancing vitamin to those you care about.

Mexican Potato Skillet Dinner

Tortillas, tomatoes, corn and olives gild the humble potato with a wealth of flavors. Top with a dollop of fat-free sour cream and a garnish of thinly sliced green onions.

4 cups cubed potatoes
2 tablespoons olive oil
1 cup chopped onion
1/2 cup diced red bell pepper
4 garlic cloves, minced
1 teaspoon ground cumin
1/2 teaspoon salt
2 cups peeled, seeded and diced
   tomatoes
1/2 cup fresh or frozen, thawed,
   corn kernels
1/4 cup chopped, pitted ripe olives

2 corn tortillas, cut into thin,
   2-inch strips
1 tablespoon minced fresh
   cilantro leaves
1/2 cup (2 oz.) shredded sharp
   Cheddar cheese
2 cups thinly sliced lettuce leaves
1/2 cup fat-free sour cream
4 green onions, thinly sliced on
   diagonal
Cilantro sprigs, for garnish

Cook the potatoes in a large pan of boiling water until just firm-tender, about 6 to 7 minutes. Drain well.

Heat the oil in a large, nonstick skillet or Dutch oven over medium-high heat. Add onion and cook, stirring occasionally, until onion begins to brown, about 5 to 6 minutes. Add potatoes and cook, stirring frequently, about 5 to 6 minutes, or until potatoes are nearly tender. Push potatoes to one side and add bell pepper, garlic, cumin and salt, stirring until combined. Stir into potato mixture along with the tomatoes, corn, olives, tortillas and cilantro. Reduce heat and cook, stirring occasionally, until the potatoes are tender, about 3 to 4 minutes. Stir in cheese and remove from heat.

To serve, mound potatoes in the center of individual plates. Surround with a ring of lettuce leaves. Top with a dollop of sour cream and a sprinkle of green onions. Garnish with cilantro sprigs.

Makes 4 servings.

Spinach Dumplings with Fresh Tomato Sauce

Striking color contrasts and mouth-watering flavor. If you don't have fresh tomatoes, substitute diced canned tomatoes.

Fresh Tomato Sauce (see opposite)
1 (10-oz.) bag fresh spinach, stems
   removed
1 cup part-skim ricotta cheese
2 eggs, beaten
2/3 cup freshly grated Parmesan
   cheese
1 cup unbleached all-purpose flour
1/8 teaspoon freshly grated nutmeg
1/4 teaspoon salt

FRESH TOMATO SAUCE
1 tablespoon olive oil
1 onion, chopped
2 garlic cloves, minced
4 cups peeled, seeded and finely
   chopped tomatoes
1 tablespoon balsamic vinegar
1 tablespoon tomato paste
1 tablespoon minced fresh parsley
1/4 teaspoon dried marjoram
1/8 teaspoon salt
Dash of freshly ground black pepper

Prepare sauce. Bring a large pot of water to a low boil.

Carefully rinse the spinach in lukewarm water to remove all the sand and grit. Shake off excess water. Place the spinach with just the water that clings to the leaves in a large, nonstick skillet over medium-high heat. Toss the spinach until it is just wilted, about 2 minutes. Place the cooked spinach in a sieve and press out excess water with the back of a spoon. Wrap spinach in paper towels and squeeze out remaining water.

Place the spinach leaves on a cutting board and finely chop. Place chopped spinach in a large bowl and stir in the ricotta. Add eggs, 1/2 cup of the Parmesan cheese, 2/3 cup of the flour, the nutmeg and salt. Stir gently until combined. Form dough into ovals the size of small eggs and dredge lightly in the remaining flour. The dough will be soft, so handle gently.

Cook dumplings in the large pot of boiling water, several at a time, until they rise to the surface, about 3 to 4 minutes. Stir gently to make sure dumplings aren't sticking to the bottom. Drain well and keep warm.

To serve, place pools of sauce on individual plates. Divide spinach dumplings among plates. Top with remaining Parmesan cheese. Serve extra sauce on the side.

Makes 4 servings.

Fresh Tomato Sauce

Heat the oil in a nonstick Dutch oven over medium heat. Add the onion and cook, stirring frequently, until the onion is tender, about 5 to 6 minutes. Add garlic and cook, stirring, 1 minute.

Add tomatoes, vinegar, tomato paste, parsley, marjoram, salt and pepper. Reduce heat to low and simmer, uncovered, 20 to 25 minutes, stirring occasionally, until the sauce begins to thicken. Serve warm.

Education

If you are using just a few tablespoons of tomato paste from a can, don't discard the rest. Just lightly oil a baking sheet, then drop the remaining tomato paste, by tablespoons, onto the sheet. Freeze until firm, then transfer the dollops of tomato paste to a self-sealing plastic bag. Freeze until needed.

Spaghetti Squash with Zucchini in Cream Sauce

A change of pace from pasta, spaghetti squash is tasty smothered in a variety of sauces, like this light, creamy herbed sauce.

1 medium spaghetti squash	1/2 cup fat-free sour cream
2 teaspoons olive oil	1/3 cup freshly grated Parmesan
1 small onion, chopped	cheese
2 cups diced zucchini	1/8 teaspoon freshly ground black
2 garlic cloves, minced	pepper
2 tablespoons minced fresh parsley	Dash of freshly grated nutmeg
1 tablespoon minced fresh basil	Tomato wedges, for garnish
1 (12-oz.) can evaporated skim milk	Parsley sprigs, for garnish

Halve the spaghetti squash lengthwise and remove seeds. Place half the squash on a plate, cut side down, and microwave, uncovered, on HIGH 7 minutes, or until tender. Keep warm. Microwave remaining half 7 minutes and keep warm.

Meanwhile, heat the oil in a large, nonstick skillet over medium heat. Add the onion and cook, stirring frequently, until the onion is translucent, about 2 to 3 minutes. Add zucchini and cook, stirring occasionally, until the zucchini is firm-tender, about 5 to 6 minutes. Add garlic, parsley and basil and cook, stirring, 1 minute.

Add evaporated milk, sour cream, Parmesan cheese, pepper and nutmeg. Reduce heat to low and cook, stirring constantly, until the sauce is bubbly and thickens. Remove from heat.

To serve, pull the spaghetti squash strands out of the squash halves with a fork. Arrange on a warmed serving dish or individual plates. Top with sauce and garnish with tomato and parsley sprigs.

Makes 6 servings.

Education

Spaghetti squash, with its slightly crunchy texture, is a pleasing alternative to pastas. Low in calories (an 8-ounce serving has only about 75 calories), it has a mild flavor that is adaptable to many dishes in which pastas might be used. If you don't have a microwave, puncture the whole squash with a knife in several places, through to the seed cavity, and boil in a large pot of water 30 to 45 minutes, or until tender. Halve lengthwise and gently remove the strands of squash from the shell with a fork.

Navy Beans Oreganata

Fresh oregano is best for this dish, though using fresh parsley and some dried oregano will work. An added bonus: It gives off a heavenly aroma as it cooks. Serve over rice and garnish with tomato wedges.

1 1/4 cups dried navy beans	1 teaspoon fresh oregano leaves or
6 cups water	1/2 teaspoon dried oregano
1 large onion, chopped	1 cup packed fresh spinach leaves,
1 large celery stalk, halved	stems removed
l large carrot, diced	1 tablespoon minced fresh parsley
3 whole garlic cloves, peeled	1/4 teaspoon salt
3 whole cloves	Dash of freshly ground black pepper
1/2 bay leaf	

Pick over beans and rinse. Place in a large saucepan with water, onion, celery and carrot. Stick each of the garlic cloves with the stem of a whole clove. Add garlic, bay leaf and oregano to the beans and bring to a boil. Reduce heat, cover and simmer 1 hour.

Rinse spinach well. Shake off excess water and dry with a paper towel. Remove celery, clove-studded garlic and bay leaf from beans. Add spinach, parsley, salt and pepper and simmer, uncovered, stirring occasionally, until the spinach is tender and the mixture is thick, about 10 to 12 minutes.

Makes 4 servings.

Education

Although most beans must be soaked beforehand to shorten cooking time, navy beans usually cook quickly enough to skip this step for the recipe above. Although some cookbooks call for the addition of baking soda when cooking beans, soda depletes B vitamins and should not be used.

Mediterranean Vegetable Sauté

Fennel highlights an assortment of vegetables complemented by olives and feta cheese. Serve over couscous or bulgur.

2 tablespoons olive oil
1 medium red onion, halved
    and thinly sliced
1 red bell pepper, thinly sliced
1 fennel bulb, trimmed and thinly
    sliced
1 small eggplant, peeled and cubed
1 zucchini, diced
2 cups peeled, seeded and diced
    tomatoes
2 garlic cloves, minced
1 teaspoon fresh oregano or thyme
    leaves or 1/4 teaspoon dried
    thyme

1/2 teaspoon fresh rosemary or
    1/4 teaspoon dried rosemary
1 2/3 cups (or 15-oz. can) cannellini
    beans (white kidney beans),
    drained
1/4 cup crumbled feta cheese
1/4 cup pimiento-stuffed olives,
    chopped
1/4 cup chopped, pitted ripe olives
Dash of freshly ground black pepper

Heat the oil in a nonstick Dutch oven over medium-high heat. Add the onion, bell pepper, fennel and eggplant and cook, stirring frequently, until the vegetables begin to brown, about 7 to 8 minutes. Add zucchini, tomatoes, garlic, oregano and rosemary and cook, stirring frequently, 3 to 4 minutes, or until the zucchini is firm-tender.

Stir in beans, feta, olives and black pepper. Reduce heat, cover and simmer 3 to 4 minutes or just until vegetables are tender.

Makes 6 servings.

Education

Fennel is slowly gaining in popularity in the United States. Its aniselike flavor imparts a special taste to the dishes in which it is used. Rich in vitamin C, with up to one-third of the RDA for vitamin A, and containing calcium, iron and magnesium, this nutritious vegetable is very low in calories, with just 23 per serving. Want to eat more fennel? Substitute this flavorful vegetable wherever you would use celery.

Thai Noodles with Cilantro

An exotic dish, but one that is easily created in your own kitchen. It is high in protein, low in fat and rich in both flavor and nutrition.

1/4 cup fresh lime juice

2 tablespoons soy sauce

1 tablespoon light brown sugar

1 tablespoon dark molasses

4 cups broccoli flowerets, in bite-size pieces

1/2 red bell pepper, thinly sliced

1/2 pound thin spaghetti, broken in half

1 tablespoon canola oil

2 teaspoons minced gingerroot

2 teaspoons minced garlic

1 teaspoon minced jalapeño chile

1 cup frozen green peas

1/4 cup thinly sliced green onions

2 tablespoons minced fresh cilantro

2 eggs, beaten

1/4 cup chopped peanuts, for garnish

Cilantro sprigs, for garnish

Combine the lime juice, soy sauce, sugar and molasses in a small bowl and set aside.

Steam the broccoli over boiling water 2 minutes. Add bell pepper and steam 2 to 3 minutes, until vegetables are crisp-tender. Run under cold water until cooled and set aside.

Bring a large pot of water to a boil and cook the spaghetti 7 to 8 minutes, or just until barely tender. Drain and set aside.

Meanwhile, heat the oil in a nonstick Dutch oven over medium heat. Add the gingerroot, garlic and chile and cook, stirring frequently, about 1 minute. Add broccoli, bell pepper, peas, green onions and cilantro and stir until heated through. Transfer the broccoli mixture to a large bowl and cover with foil to keep warm.

Pour the beaten eggs into the Dutch oven over medium heat, and as they begin to cook, add cooked spaghetti and lime juice mixture. Using two forks, toss the mixture until the liquid is absorbed. Add broccoli mixture and toss until thoroughly mixed. Serve on a warmed platter or individual plates. Garnish with peanuts and cilantro sprigs.

Makes 4 servings.

Education

Raw eggs carry the risk of salmonella, a food poisoning that can be serious for the very young, the elderly or anyone with a compromised immune system. In the recipe above, the eggs are completely cooked as the pasta is tossed in the pan over heat. Be careful with egg dishes that do not result in hard cooking of both whites and yolks.

Broccoli Ravioli with Fresh Tomatoes

Wonton wrappers make these ravioli easy to stuff with a tasty broccoli filling. If tomatoes aren't in season, drizzle your favorite tomato pasta sauce, thinned with tomato juice, over the ravioli.

1 large stalk broccoli
2 teaspoons olive oil
1 onion, finely chopped
1 garlic clove, minced
1 teaspoon fresh marjoram leaves
    or 1/2 teaspoon dried
    marjoram
Dash of freshly grated nutmeg
1/3 cup (about 1 1/2 oz.) shredded
    part-skim mozzarella cheese
1/3 cup fat-free cream cheese
1 (16-oz.) package 3 1/2-inch-square
    wonton wrappers

1 tablespoon unbleached all-purpose
    flour
2 tablespoons water
4 large tomatoes, peeled, seeded,
    chopped and drained
1 green onion, finely chopped
4 basil leaves, thinly sliced
1/2 teaspoon sugar
1/8 teaspoon salt
Dash of freshly ground black pepper
1/4 cup freshly grated Parmesan
    cheese

Bring a large pot of salted water to a boil.

Peel the broccoli stem and finely chop stem and top. Heat the oil in a medium, nonstick skillet over medium heat. Add the onion and cook, stirring occasionally, until the onion is translucent, about 2 to 3 minutes. Add broccoli, garlic, marjoram and nutmeg. Cook, stirring, until the broccoli is tender, about 5 to 6 minutes.

Combine the mozzarella and cream cheeses in a medium bowl, spreading the mixture out across the bottom and sides of bowl. Add hot broccoli mixture and stir until well combined.

Separate 32 wonton wrappers. Keep in plastic wrap so wrappers do not dry out as you work. To make ravioli, place a wonton wrapper on a clean work surface and place about 1 1/2 tablespoons of the broccoli filling in the center. Flatten the filling slightly.

Stir together flour and water in a small dish. Brush the mixture along the outside edge of the wrapper. Top ravioli with another wrapper, lifting the edges upward to align and pinch to seal. Set aside, covered with a moist paper towel, until all are made. Repeat with remaining filling and wrappers.

Cook half the ravioli in the pot of boiling water 4 to 5 minutes, or until ravioli are tender. Turn once to make sure both sides are cooked. (Handle carefully so that wrappers are not punctured.) Drain and keep warm while cooking remaining ravioli.

Meanwhile, combine chopped tomatoes, green onion, basil, sugar, salt and pepper in a medium saucepan over medium heat and bring to a boil. Reduce heat, cover and simmer 2 minutes.

Arrange 3 or 4 ravioli on individual plates. Top with tomatoes and sprinkle with Parmesan cheese.

Makes 4 servings.

Zucchini Fettuccine

Not only does this sound nifty, but it is! The zucchini are cut into long, thin strips, boiled with pasta and peas and tossed with pesto. Garnish the plate with vibrant red, garden-ripe tomatoes and sprigs of basil. This meal is a celebration of summer.

2 medium zucchini	1/2 cup fresh or frozen green peas
Salt	1/2 cup Pesto Genovese (page 61)
1/4 pound fettuccine, broken in half	

Trim stem and blossom ends from zucchini. Cut lengthwise into 1/4-inch-thick slices. Lay the slices flat and cut each into long, thin strips. Place in a large bowl and sprinkle liberally with salt. Let stand 30 minutes. Rinse zucchini thoroughly.

Meanwhile, bring a large pot of water to a boil. Add fettuccine and cook, uncovered, 12 to 13 minutes, or until quite firm-tender. If using fresh peas, add them after pasta has boiled 5 minutes. If using frozen peas, add with zucchini at the end of the cooking time and return to a boil.

Drain the vegetables and pasta immediately. Do not let the pasta or zucchini become overcooked.

Spread half the pesto in the bottom of a warmed serving bowl. Add hot pasta mixture. Top with remaining pesto and toss until well combined.

Makes 4 servings.

Education

To get the beta carotene from summer squashes, you must eat them with their skins. The inner pulp has no beta carotene. If you grow zucchini and summer squash in the garden, pick them before they get too large and the skins are too tough to enjoy.

Vegetable Macaroni & Lentil Goulash

This looks like a ground meat dish we used to enjoy, but it has no animal protein or animal fats. The delicious taste and appearance appeal even to my "don't-try-anything-new" teenager.

1/2 cup red lentils (see Note, page 124), picked over and rinsed
1/4 cup brown lentils, picked over and rinsed
2 tablespoons olive oil
1 large onion, chopped
2 large carrots, peeled and finely diced

1/4 teaspoon dried marjoram
1/4 teaspoon salt
1/8 teaspoon freshly ground black pepper
2 cups water
1 (28-oz.) can whole tomatoes, drained and chopped
1 (8-oz.) can tomato paste
3/4 pound macaroni, cooked

Pick over and rinse red and green or brown lentils. Set aside.

Heat the oil in a nonstick Dutch oven over medium heat. Add the onion and carrots and cook, stirring frequently, until the onion is tender, about 5 minutes. Add the lentils, marjoram, salt and pepper, and cook, stirring, 2 to 3 minutes.

Add water and bring to a boil. Reduce heat, cover and simmer 15 minutes, until the lentils are firm-tender and the water is absorbed. Stir in tomatoes and tomato paste and bring to a boil. Reduce heat, cover and simmer 5 minutes. Add drained cooked macaroni and heat through before serving.

Makes 6 servings.

Education

Lentils are a good source of saponin, a phytochemical believed to lower blood pressure and cholesterol levels. Of course, substituting lentils for meat eliminates any saturated fat from that source, another plus when cholesterol levels are a concern.

Green Bean Spaghetti

A vegetarian dish with a twist on beans—this time fresh green beans instead of reconstituted dried beans, a favorite of friends Lydia and Bob. Serve with a big salad and French bread.

1 pound green beans	1 tablespoon minced fresh basil or
1 tablespoon olive oil	1/2 teaspoon dried
4 garlic cloves, minced	1 teaspoon minced fresh marjoram
1 1/4 pounds thin spaghetti	or 1/4 teaspoon dried
6 cups Herbed Tomato Sauce	1/8 teaspoon freshly ground black
(page 149) or bottled sauce	pepper

Trim ends from green beans, leaving beans whole. Heat the oil in a nonstick Dutch oven over medium heat. Add the beans and garlic and cook, stirring frequently, until tender, about 15 minutes.

Bring a large pot of water to a boil. Add spaghetti and cook until firm-tender. Drain.

Meanwhile, add spaghetti sauce, basil, marjoram and pepper to beans. Heat through. Serve over pasta.

Makes 6 servings.

Education

Green beans and dried beans are two ends of one spectrum. Dried beans result when green beans are left to mature on the stalk. The seeds grow to full size and dry in the pod. The nutritional difference for you is that the immature pods, eaten fresh, are a good source of beta carotene and vitamin C, while protein is the main contribution of dried beans.

Herbed Tomato Sauce

If you don't have fresh tomatoes, use 4 cups of canned Italian tomatoes with their juice. Also, try the variations listed below.

2 tablespoons olive oil
1 red bell pepper, diced
1 large onion, chopped
3 garlic cloves, minced
6 large, garden-ripe tomatoes,
    peeled, seeded and diced
1/3 cup tomato paste
2 tablespoons minced fresh parsley
2 teaspoons minced fresh basil or
    1/2 teaspoon dried basil

1 teaspoon fresh oregano or
    1/4 teaspoon dried oregano
1 teaspoon fresh marjoram or
    1/4 teaspoon dried marjoram
1/2 teaspoon salt
1/8 teaspoon freshly ground black
    pepper

Heat the oil in a large, nonstick Dutch oven over medium heat. Add the bell pepper and onion and cook, stirring occasionally, until the onion is translucent, about 3 to 4 minutes. Add garlic and stir 1 minute more.

Add tomatoes, tomato paste, parsley, basil, oregano, marjoram, salt and black pepper. Stir well and bring to a boil. Reduce heat, cover and simmer, stirring occasionally, 30 to 40 minutes, or until sauce is thickened.

Makes about 4 cups.

Variations

Herbed Tomato Sauce II Add 3/4 cup shredded carrots with the garlic. Stir until carrots are wilted. Add remaining ingredients and continue as above.

Mushroom Tomato Sauce Add 2 cups sliced mushrooms with the garlic. Cook, stirring occasionally, until the mushrooms are tender and the liquid has evaporated. Add remaining ingredients and continue as above.

Linguine with Garlic & Pine Nuts Marinara

A similar sauce, discovered while on vacation, became the basis for this delicious creation. Guests seem to love it. Serve with a large salad and garlic bread.

1 tablespoon olive oil
1 small onion, chopped
6 to 8 garlic cloves, minced
1 (28-oz.) can Italian-style plum
    tomatoes
2 fresh tomatoes, cored and
    quartered
1 (15-oz.) can small ripe olives,
    drained and chopped
1/2 cup pine nuts
1 tablespoon fresh basil leaves,
    thinly sliced, or 1 teaspoon
    dried basil, crumbled

1 teaspoon chopped fresh oregano
    leaves or 1/4 teaspoon dried
    oregano
1/2 teaspoon salt
1/4 teaspoon freshly ground black
    pepper
1 1/4 pounds linguine or spaghetti,
    cooked

Heat oil in a large skillet over medium-low heat. Add onion and garlic and cook, stirring occasionally, until onion is translucent, about 3 to 4 minutes.

Meanwhile, drain tomatoes, reserving juice. Add canned and fresh tomatoes to a food processor and process until pureed.

Add pureed tomatoes, reserved tomato juice, olives, pine nuts, basil, oregano, salt and pepper to skillet and bring to a boil. Reduce heat, cover and simmer 15 to 20 minutes, or until slightly thickened and onion is tender.

Toss sauce with pasta and serve on warmed plates.

Makes 6 servings.

Salads & Sandwiches

About Salads

Salads are a wonderful way to enjoy raw vegetables. Fresh greens can provide interesting textural contrast when everything else in the meal is cooked, mashed or pureed. And even as salads become more popular, they are growing more varied. Some are even main-dish—a meal in a bowl.

Iceberg, the nutritionally poor cousin of today's wide selection of salad greens, is being replaced by romaine and red leaf lettuces as well as by specialty "leaves," like arugula, watercress and endive. Remember, the darker the greens, in general, the more nutritious they are.

Little things can make a big difference in a salad: using fresh lemon juice, freshly ground pepper, fresh herbs, a good quality oil. Also, if you want the salad dressing to flavor the salad, and not just sink dispiritedly to the bottom of the bowl, the lettuce and other ingredients must be dry.

One of the best "salad" investments I've made is a spinning basket that allows me to rinse and dry greens in a minute. Making salads convenient means you'll include them in your diet regularly. Leftover greens from salads can be refrigerated in a self-sealing plastic bag and remain crisp for the next day as long as you haven't used dressing. So serve dressing "on the side," so each person can spoon a small amount on the individual salad.

Salad Greens

Don't be loyal to just one lettuce when you are making salads. Play the field!

There are many types of greens available in your supermarket, each with a unique contribution to make in terms of flavor and nutrition. Be a matchmaker, and see which combinations of leaves bowl you over.

Arugula Very rich in beta carotene, these mustardlike greens also contain more vitamin C than most other greens. Arugula is a cruciferous vegetable, thought to protect against cancer.

Belgian endive While not outstanding in the nutrient department, this green adds a crisp, slightly bitter, but flavorful, touch to salads.

Chicory (curly endive) It is about equal to Romaine in beta carotene and vitamin C content, but richer in calcium, and it has an interesting bitter flavor.

Romaine Its deep-green leaves are crisp, like its pale cousin, iceberg, but more nutritious.

Watercress Another member of the cruciferous family of greens, it is richer in beta carotene and vitamin C than nearly all other greens except arugula.

Vegetarian Sandwiches

Sandwiches are made-to-order for vegetarian complementary proteins. Rolls, pita bread and multigrain slices provide grain proteins to augment beans and vegetables. Attractive and portable, sandwiches fit easily into today's busy lifestyle.

Many of the sandwich fillings here could also be served as a main dish. For example, Veggie Burgers, Curried Red Lentil Burgers and Black Bean Burgers could take their proper place on a dinner plate. In the same way, Falafel, in the Main Dish chapter, would also be just wonderful if served in pita bread, drizzled with tahini sauce and topped with tomatoes.

Creamy French Salad Dressing

This dressing gives just the right piquant touch to a green salad.

1/3 cup extra-virgin olive oil
1/3 cup low-fat yogurt
1/4 cup tomato juice
3 tablespoons rice vinegar
1 teaspoon grated onion

1 garlic clove, minced
1/4 teaspoon salt
1/4 teaspoon ground yellow
    mustard
1/4 teaspoon paprika

Combine ingredients in a blender or food processor and process until smooth. Keep refrigerated.

Makes 1 cup.

Cleopatra Dressing

A Caesar-style dressing, bursting with flavor, but without the anchovies . . . an excellent dressing on tossed green salads.

1 tablespoon low-fat yogurt
1 teaspoon sesame tahini
1 1/2 tablespoons water
1 1/2 tablespoons fresh lemon juice
1/4 cup extra-virgin olive oil

1 garlic clove, minced
1/2 teaspoon Dijon mustard
Dash of salt
Dash of freshly ground pepper

Cream together a little of the yogurt and the tahini in a small bowl. Gradually add remaining yogurt and water and stir until smooth. Add lemon juice, olive oil, garlic, mustard, salt and pepper and whisk with a fork until combined.

Makes about 1/2 cup.

Antipasto Pasta Salad

Flavorful, substantial—a good choice when you're looking for something special.

1/2 pound fusili or spiral pasta
1 pound fresh green beans,
    trimmed
1/2 small red onion, thinly sliced
2 tomatoes, peeled, seeded and
    diced
1 small head red leaf lettuce, torn
    into bite-size pieces

1 cup cooked chickpeas
1 (8 1/2-oz.) can artichoke hearts,
    drained and quartered
1 cup pitted ripe olives, halved
    lengthwise
2 tablespoons chopped green olives
Cleopatra Salad Dressing (see
    page 153) or oil and vinegar

Bring a large pot of salted water to a boil over medium-high heat. Add pasta and stir. Return to a boil and cook 8 to 10 minutes, or until firm-tender. Drain in a colander and run under cold water until cooled. Drain and place in a large serving bowl.

While cooking pasta, bring a large saucepan of water to a boil. Add green beans and cook 10 to 15 minutes, or just until crisp-tender. Run under cold water until cooled. Drain and place in serving bowl with pasta.

Separate onion slices into rings and add to bowl. Add tomatoes, lettuce, chickpeas, artichoke hearts and olives and toss until combined. Dress with Cleopatra Salad Dressing or oil and vinegar.

Makes 8 servings.

Variation

Substitute 1 (6-oz.) jar marinated artichoke hearts, drained, rinsed and quartered, for canned artichoke hearts.

Julienne Vegetable Salad

Light steaming enhances a colorful mixed vegetable salad. You'll find mild rice vinegar at Oriental food stores.

1 large carrot, cut into 2 × 1/4-inch
    strips
1 small zucchini, cut into
    2 × 1/4-inch strips
2 celery stalks, cut into
    2 × 1/4-inch strips
1 cup fresh or frozen green peas

DILL DRESSING
2 tablespoons extra-virgin olive oil
3 tablespoons rice vinegar
1/2 teaspoon minced fresh dill weed
1/4 teaspoon Dijon mustard
1/4 teaspoon salt
Dash of freshly ground black
    pepper

Cook the carrot strips in a steaming basket over boiling water just until crisp-tender, about 3 to 4 minutes. Run under cold water until cool and place in a serving bowl. Steam the zucchini just until firm-tender, about 2 to 3 minutes. Run under cold water until cool and add to carrots. Steam the celery just until crisp-tender, about 2 to 3 minutes. Run under cold water until cool and add to serving bowl. Steam the peas just until firm-tender, 1 to 2 minutes for frozen peas, 4 to 5 minutes for fresh. Run under cold water until cool and add to other vegetables.

Prepare dressing. Pour over vegetables, toss and serve at room temperature.

Makes 4 servings.

Dill Dressing

Whisk together oil, vinegar, dill, mustard, salt and pepper in a small bowl.

Makes about 1/4 cup.

Variation

If you are using a more acidic vinegar, substitute half water and half vinegar for the rice vinegar.

Avocado Citrus Salad with Creamy Dressing

A great, if nontraditional, combination of tastes and textures. The dressing is terrific on any fruit salad.

1 pink grapefruit, peeled, seeded
    and sectioned
1/4 medium cantaloupe, peeled
    and cubed
1 cup red seedless grapes
1 avocado, peeled, pitted, quartered
    and sliced
1/4 cup diced celery
1/4 cup lightly toasted walnuts
Creamy Dressing (opposite)
Spinach leaves, for garnish

CREAMY DRESSING
3/4 cup low-fat yogurt
2 tablespoons lime juice
2 tablespoons honey
1/8 teaspoon ground coriander
1/8 teaspoon ground cardamom
Dash of salt

Halve or quarter grapefruit sections and place in a large bowl with cantaloupe, grapes, avocado, celery and walnuts. Toss to combine.

Prepare dressing. Line individual salad plates with spinach. Spoon fruit salad over spinach. Serve dressing on the side.

Makes 4 servings.

Creamy Dressing

Combine all ingredients in a blender or food processor and process until smooth.

Makes about 1 cup.

Education

Avocados are an amazing source of potassium; ounce-for-ounce, they provide 60 percent more potassium than that found in bananas. Avocados are also rich in beta carotene and folacin, and have enough vitamin C that they were once called "midshipmen's butter." These green fruits were taken on long ocean voyages to prevent scurvy.

Their only drawback is that they are high in fat, although the fat is mostly monounsaturated (like that found in olive oil), which has been shown to lower blood cholesterol. To balance this, when you serve avocado in salad, add a very low-fat or fat-free dressing.

Mary's Mandarin Spinach & Onion Salad

Canned mandarin oranges make this a refreshing citrus salad for any time of year.

1 (10-oz.) package fresh spinach
    leaves
1 (6 1/2-oz.) can mandarin oranges
1 cup sliced mushrooms
1/2 small red onion, thinly sliced
    and separated into rings

ORANGE-HONEY DRESSING
1/4 cup fresh orange juice
3 tablespoons extra-virgin olive oil
1 tablespoon honey
1 tablespoon fresh lemon juice
1/4 teaspoon paprika
Dash of salt

Wash the spinach well in several changes of water to remove all the grit. Dry leaves well. Remove stems and tear leaves into large, bite-size pieces. Drain orange sections. Combine spinach, orange sections, mushrooms and onion rings in a large salad bowl.
    Prepare dressing. Pour dressing over salad and toss just before serving.

Makes 6 servings.

Orange-Honey Dressing

Combine all ingredients in a small bowl or blender. Stir or process until smooth.

Makes about 1/2 cup.

Marinated Lentil Salad with Herb Dressing

The dressing is thick with herbs! Take notice of the wonderful aromas wafting from the bowl when you pour the dressing over hot lentils . . .

1 cup dried brown or green lentils
2 cups water
3 garlic cloves, minced
1 shallot, minced
1/2 teaspoon salt
Herb Dressing (opposite)
1 carrot, finely diced
1 small celery stalk, thinly sliced
1/2 cup peeled, seeded and diced
    tomato
2 green onions, thinly sliced on the
    diagonal
1/4 cup chopped ripe olives

HERB DRESSING
3 tablespoons extra-virgin olive oil
1 tablespoon fresh lemon juice
1 tablespoon balsamic vinegar
1 tablespoon minced fresh parsley
2 teaspoons minced fresh basil
2 teaspoons minced fresh chives
1 teaspoon minced fresh dill weed
1 teaspoon Dijon mustard
1/4 teaspoon salt
1/8 teaspoon freshly ground black
    pepper
2 garlic cloves, minced

Bring the lentils, water, garlic, shallot and salt to a boil in a medium saucepan. Reduce heat, cover and simmer 15 minutes, or until lentils are firm-tender, not mushy. Drain off any excess water. Place lentils in a serving bowl.

Prepare dressing. Pour dressing over the hot lentils and toss until combined.

When lentils are cool, add the carrot, celery, tomato, green onions and olives. Toss well. Serve at room temperature or chilled.

Makes 6 side-dish servings or 4 main-dish servings.

Herb Dressing

Whisk all ingredients together in a small bowl.

Makes about 1/3 cup.

Variations

Use marinated lentils alone as a salad, or toss lentils with 2 tablespoons minced pimiento and 1/2 cup cooked whole-kernel corn and serve on lettuce leaves.

Education

Lentils are an important source of protein throughout the Middle East, India and Nepal. They are an especially good source of folacin, which helps our bodies build new cells. And, fortunately, lentils taste as good in salads as they do in soups. In addition to the salad here, flavorful marinated lentils can be sprinkled over green salads, too.

Macaroni & Tomato Salad for One

A gorgeous salad! When you're eating alone, you'll still feel special and it's so very easy to make.

1 cup cooked macaroni
2 green onions, thinly sliced
1 celery stalk, finely diced
1 tablespoon Cleopatra Dressing
   (page 153)

1 large, garden-ripe tomato, cut
   into 8 wedges
1 small Romaine lettuce leaf
1 parsley sprig, for garnish

Combine the macaroni, green onions and celery in a large bowl, and toss with the dressing until combined.

Arrange tomato wedges over the salad. Garnish with lettuce leaf and parsley.

Makes 1 main-dish serving.

Black Bean Salad on Spinach

Colorful vegetables and beans make a quick and attractive salad, pretty enough for a party!

1 2/3 cups cooked (or 15-oz. can)
    black beans, drained
1 (8 1/2-oz.) can whole-kernel
    corn, drained
1/2 red bell pepper, finely diced
3 radicchio leaves, thinly sliced
2 green onions, thinly sliced on
    the diagonal
1/4 cup finely diced celery
1 tablespoon minced fresh parsley
Lemon-Soy Dressing (opposite)
Spinach leaves, stems removed
8 thin slices red onion, for garnish
Parsley sprigs, for garnish

LEMON-SOY DRESSING
2 tablespoons extra-virgin olive oil
1 tablespoon fresh lemon juice
1 teaspoon soy sauce
1/2 teaspoon Dijon mustard
1/2 teaspoon light brown sugar
Dash of freshly ground black pepper

Combine the beans, corn, bell pepper, radicchio, green onions, celery and minced parsley in a medium bowl.

Prepare dressing. Pour dressing over bean mixture and toss. Arrange spinach leaves on serving plates. Top with bean salad and garnish with onion and parsley sprigs.

Makes 4 servings.

Lemon-Soy Dressing

Combine oil, lemon juice, soy sauce, mustard, sugar and pepper in a small bowl and whisk until combined.

Makes about 1/4 cup.

Creamy Carrot & Raisin Salad

Sweet, with a touch of citrus! A creamy dressing without the cream.

2 cups grated carrots
1 small celery stalk, finely diced
1/2 Granny Smith apple, peeled,
    cored and finely diced
1/2 cup raisins
1/4 cup chopped pecans or walnuts
Orange-Yogurt Dressing (opposite)
1 small head butter lettuce

ORANGE-YOGURT DRESSING
1/2 cup low-fat yogurt
1/2 cup fresh orange juice
2 tablespoons extra-virgin olive oil
2 teaspoons honey
1 teaspoon fresh lemon juice
1/2 teaspoon ground coriander
1/8 teaspoon paprika
1/8 teaspoon salt

Combine the carrots, celery, apple, raisins and pecans or walnuts in a serving bowl. Prepare dressing. Pour dressing over carrot mixture and toss until combined. Arrange lettuce leaves on salad plates and top with a scoop of carrot mixture.

Makes 4 servings.

Orange-Yogurt Dressing

Combine all ingredients in a blender or food processor and process until smooth.

Makes about 1 1/4 cups.

Hot Potato Salad

Herbs meld the flavors of potatoes, onion and green peas for a mouth-watering hot salad for cool days.

2 pounds small red potatoes	1 tablespoon water
1/4 cup chopped red onion	3 tablespoons minced fresh parsley
1 cup frozen green peas	1 tablespoon fresh basil leaves
	1 teaspoon fresh tarragon leaves
HERB-YOGURT DRESSING	1 teaspoon Dijon mustard
1/3 cup low-fat yogurt	1/4 teaspoon salt
1/4 cup fat-free sour cream	1/8 teaspoon freshly ground black
3 tablespoons olive oil	pepper
3 tablespoons lemon juice	

Scrub and cut potatoes into large, bite-size cubes. Place in a large saucepan and cover with salted water. Bring to a boil, reduce heat and simmer 15 to 20 minutes, or until firm-tender. Add onion and cook 1 minute. Drain.

Meanwhile, place peas in a small, microwave-safe covered casserole dish. Microwave on HIGH 3 minutes.

Prepare dressing. Place the hot potatoes, onion and peas in a large serving bowl. Pour dressing over salad and toss until vegetables are coated with dressing. Serve immediately.

Makes 4 servings.

Herb-Yogurt Dressing

Combine the yogurt, sour cream, oil, lemon juice, water, parsley, basil, tarragon, mustard, salt and pepper in a blender or food processor and process until smooth.

Makes about 3/4 cup.

Variation

Chill salad before serving. If salad will be chilled, refresh peas under cold water after cooking to cool. Let potato and dressing mixture cool before adding peas. Toss to combine.

Green-on-Green Tossed Salad

Assemble several greens when creating a tossed salad. The texture and flavor will increase dramatically!

1 head red leaf lettuce
1/2 cup arugula or watercress
    leaves
2 cups thinly sliced spinach
    leaves
1/2 cup chopped cabbage

1/2 cup thinly sliced celery
1/4 cup finely diced green bell
    pepper
Cleopatra Dressing (page 153) or
    Creamy French Salad
    Dressing (page 153)

Rinse and dry lettuce and tear into large, bite-size pieces. Rinse and dry the arugula or watercress and tear into bite-size pieces. Combine lettuce, arugula or watercress, spinach, cabbage, celery and bell pepper in a salad bowl. Add dressing, and toss to combine just before serving.

Makes 4 servings.

Variations

Use a small head of Romaine lettuce in place of red leaf; substitute chicory or chopped broccoli rabe (rapini) for arugula; use thinly sliced fennel bulb or Belgian endive in place of celery.

Caribbean Cabbage Salad

A refreshing and simple salad. Lime juice and hot pepper sauce give it tang.

3 cups finely chopped cabbage
1 tomato, peeled, seeded and
    chopped
1 celery stalk, thinly sliced on the
    diagonal
2 tablespoons fresh lime juice

1 1/2 tablespoons extra-virgin
    olive oil
1/4 teaspoon salt
Dash of freshly ground black
    pepper
Dash of hot red pepper sauce

Combine the cabbage, tomato and celery in a serving bowl. In a small bowl, combine lime juice, oil, salt, pepper and pepper sauce to taste. Pour mixture over cabbage mixture and toss until combined. Chill before serving.

Makes 4 servings.

Five-Vegetable Tabbouleh

Friends find the crunch of zucchini a nice change from the softer cucumber that is traditional with this Middle Eastern salad.

3/4 cup bulgur
1 1/2 cups boiling water
1/2 cup finely diced celery
1/2 cup finely diced zucchini
1 carrot, finely diced
2 tomatoes, peeled, seeded and
    diced

2 green onions, thinly sliced on the
    diagonal
3 tablespoons minced fresh parsley
1 lemon
2 tablespoons extra-virgin olive oil
1/4 teaspoon salt
Dash of freshly ground pepper

Place the bulgur in a medium bowl and cover with boiling water. Let stand 15 minutes, or until grains are plump. Drain. Combine bulgur, celery, zucchini, carrot, tomatoes, green onions and parsley in a serving bowl. Grate some of the lemon peel over the salad, to taste.

Halve and juice lemon. Combine 2 tablespoons lemon juice with oil, salt and pepper in a small bowl. Pour over the salad and toss until combined.

Makes 2 main-dish servings.

Broccoli, Tomato & Onion Salad

Easy to make, this salad needs to be prepared just a day ahead.

4 stalks broccoli	1 tablespoon water
3 or 4 medium, garden-ripe	1/2 garlic clove, minced
tomatoes	1/2 teaspoon salt
1 small red onion	1/4 teaspoon dried oregano,
	crumbled
OIL & VINEGAR DRESSING	1/4 teaspoon dried marjoram,
1/4 cup extra-virgin olive oil	crumbled
2 tablespoons cider vinegar	Dash of freshly ground pepper

Peel broccoli stalks and thinly slice. Separate broccoli into bite-size pieces. Place in a large serving bowl.

Cut tomatoes into wedges. Dice the onion. Add tomatoes and onion to broccoli.

Prepare dressing. Add dressing to salad and toss well to combine. Cover and refrigerate overnight, tossing once or twice.

Makes 6 servings.

Oil & Vinegar Dressing

Whisk the dressing ingredients together in a small bowl.

Taco Salad Picnic Buffet

Arrange chips and toppings on the table and let everyone choose their own. The combination of hot, cooked chili and cold, crunchy vegetables heightens the appeal of this salad.

1 large carrot, cut into pieces
1 small onion, quartered
1 tablespoon extra-virgin olive oil
2 large garlic cloves, minced
1 teaspoon chili powder
1 teaspoon ground cumin
1 teaspoon ground coriander
1/4 teaspoon dried oregano, crumbled
1 2/3 cup cooked (or 15-oz. can)
    black beans, drained
1 2/3 cup cooked (or 15-oz. can)
    red kidney beans, drained
1 (14 1/2 oz.) can Mexican-style
    stewed tomatoes

TOPPINGS
Low-fat tortilla chips
Coarsely shredded lettuce
Thinly sliced celery
Chopped fresh broccoli
Chopped fresh cauliflower
Diced tomatoes
Diced red onions
Thinly sliced green onions
Chopped ripe olives
Salsa
Green Onion Guacamole
   (page 4)

Place the carrot in a food processor and process until coarsely chopped. Add onion and continue to process until vegetables are finely chopped.

Heat oil in a medium saucepan over medium heat. Add carrot mixture and cook, stirring often, until mixture is slightly tender, about 3 to 4 minutes. Add garlic and cook, stirring, 1 minute. Add chili powder, cumin, coriander and oregano and stir until well combined.

Puree stewed tomatoes in a blender or food processor. Add to pan with beans and stir to combine. Bring to a boil and stir. Reduce heat, cover and simmer 10 minutes, stirring occasionally.

To assemble salads, have guests line soup-size bowls with tortilla chips and spoon on hot bean chili, then select desired toppings.

Makes 4 servings.

Variation

Instead of the combination here, you can substitute your favorite chili.

Veggie Burgers with Cashews

Cashews add a meaty flavor to delicious vegetarian burgers. Eliminate salt if you are using salted nuts.

1 large carrot, cut into large pieces
1 celery stalk, cut into large pieces
1/2 cup cashews
1/2 cup fresh bread crumbs
1 egg
1/4 cup grated onion
1 tablespoon minced fresh parsley
1/4 teaspoon salt

1/4 teaspoon ground coriander
1/2 cup stone-ground yellow
  cornmeal
About 1 tablespoon vegetable oil
4 whole-wheat rolls
4 tomato slices
1/4 cup thinly sliced green onions
1/2 cup alfalfa sprouts

Place the carrot and celery in a food processor and process until coarsely chopped. Add nuts and process until finely chopped. Add bread crumbs and egg.

Squeeze moisture from grated onion and add with parsley, salt and coriander to food processor. Process until ingredients are finely chopped and well combined, scraping down the sides of the bowl, as necessary.

Form vegetable mixture into 4 patties. Dredge patties in cornmeal. Heat oil in a nonstick skillet over medium heat. Add patties and cook, turning to brown both sides, about 10 minutes.

Serve on rolls with tomato slices. Sprinkle with green onions and top with sprouts.

Makes 4 servings.

Variation

Dilled Veggie Burgers Add 2 teaspoons minced fresh dill weed in addition to parsley.

Turkish Pita Pockets

Travelers who visit Turkey in the autumn enjoy delicious vegetables and melons. This salad-stuffed pita is reminiscent of the flat-bread sandwiches we once made there. Buy good-quality plain green olives at a Middle Eastern food store, if you can.

1 cup cooked white beans, drained
1 cup peeled, seeded and diced
    tomatoes
1 cup peeled, seeded and diced
    cucumber
1/4 cup chopped sweet Spanish
    onion or green onions
1 1/2 tablespoons extra-virgin
    olive oil
1 tablespoon fresh lemon juice

1 teaspoon minced green olives
1 teaspoon minced fresh parsley
Dash of freshly ground black
    pepper
4 pita breads, halved
Lettuce or spinach leaves
Red grapes, for garnish
Fresh melon slices, for garnish
Mint sprigs, for garnish

Combine the beans, tomatoes, cucumber and onion in a medium bowl. Whisk together the oil, lemon juice, olives, parsley and pepper in a small bowl. Pour dressing over vegetables and toss to combine.

To assemble sandwiches, open pita bread halves and line with some of the lettuce or spinach. Spoon bean mixture into pita bread. Serve with grapes, slices of fresh melon and mint sprigs.

Makes 4 servings.

Open-Face Avocado Sandwich

Creamy Herb Dressing tops an artfully arranged selection of attractive vegetables on thin, multigrain bread.

4 thin slices dark, multigrain bread
4 slices tomato
1 avocado, peeled, halved and pitted
1 thin slice red onion
1 cup alfalfa sprouts, for garnish
4 small bunches seedless red grapes,
    for garnish

CREAMY HERB DRESSING
1/4 cup low-fat yogurt
1 tablespoon fresh dill weed or
    basil leaves
1 tablespoon extra-virgin olive oil
1 garlic clove, minced
1/4 teaspoon salt
Dash of freshly ground black
    pepper

Place bread slices on a serving plate or on individual plates. Top each with a tomato slice. Cut avocado halves in half again lengthwise. Thinly slice each quarter and fan out over the tomato slices. Separate the onion slice into rings and arrange over the avocado.

Prepare dressing. Spoon a ribbon of dressing over each sandwich. Garnish with alfalfa sprouts and grapes.

Makes 4 servings.

Creamy Herb Dressing

Combine all ingredients in a food processor or blender and process until smooth.

Makes about 1/4 cup.

Grilled Vegetable Hero Sandwiches

Use a hero roll (also called a "submarine" or "hoagie" roll) to hold the assortment of grilled vegetables.

1 large (1 1/4- to 1 1/2-lb.) eggplant	1 pound mushrooms
2 small zucchini	1 cup Zesty Garlic Barbecue
1 large, sweet Spanish or red onion	Sauce (opposite), blended
2 red bell peppers	until smooth
2 large tomatoes	4 hero or hoagie rolls

Preheat charcoal grill. Trim ends from eggplant and cut lengthwise into 8 equal slices. Trim zucchini and cut in half crosswise, then cut lengthwise into 1/4-inch-thick slices. Peel onion and cut into 1/4-inch-thick slices. Core and seed bell peppers and cut lengthwise into 8 pieces. Cut tomatoes into thick slices. Trim mushroom stems.

Arrange vegetables on grill over a slow, steady heat. Turn and rearrange vegetables until they are lightly browned. Brush vegetables with barbecue sauce and continue to grill until vegetables are tender. Cut larger vegetables into thin strips and toss to combine on a large serving plate.

To serve, cut open rolls and pile high with grilled vegetables.

Makes 4 servings.

Zesty Garlic Barbecue Sauce

Brush onto eggplant, zucchini, potatoes or other grilled vegetables. Can also be used to coat and bake vegetables.

1 large onion, finely chopped
1 tablespoon canola oil
3 large garlic cloves
1 tablespoon packed light brown
    sugar
1 teaspoon ground cumin
1 teaspoon ground coriander
1/4 teaspoon dried oregano,
    crumbled

1/2 cup ketchup
1 tablespoon molasses
1/2 teaspoon salt
1/4 teaspoon hot pepper sauce
1/8 teaspoon freshly ground black
    pepper

Combine the onion and oil in a medium microwave-safe bowl. Microwave on HIGH 2 minutes.

Mince garlic and stir into onion. Microwave on HIGH 1 minute. Stir in sugar, cumin, coriander and oregano. Microwave on HIGH 1 minute.

Stir in ketchup, molasses, salt, hot sauce and pepper. Blend until smooth, if desired. Brush on grilled vegetables about 10 minutes before they are done. Store tightly covered in the refrigerator up to 1 week.

Makes about 1 cup.

Curried Red Lentil Burgers

Quick-cooking red lentils make this an easy summertime meal. Serve with soup and salad.

1/3 cup red lentils
1 1/2 cups water
1/3 cup couscous
1/2 cup diced carrot
1/4 cup grated onion, squeezed dry
About 2 tablespoons extra-virgin
    olive oil
2 teaspoons Madras curry powder
1/2 teaspoon ground coriander
1/4 teaspoon salt

1 garlic clove, minced
Dash of ground cayenne
1 tomato, peeled, seeded and finely
    diced
1/2 small cucumber, peeled, seeded
    and finely chopped
1 teaspoon minced fresh cilantro
1 fresh mint leaf, minced
2 whole-wheat pita breads, halved

Combine the lentils and 1 cup of the water in a medium bowl. Microwave on HIGH 2 1/2 to 3 minutes. Cover and let stand 10 to 15 minutes, or until firm-tender.

Meanwhile, bring the remaining 1/2 cup water to a boil in a small saucepan over high heat. Stir in couscous, remove from heat and cover. Let stand 5 minutes.

Drain the lentils. Place lentils, couscous, carrot, onion, 1 tablespoon of the oil, curry powder, coriander, salt, garlic and cayenne in a food processor. Process until smooth, scraping down sides of the bowl as necessary.

Form lentil mixture into 4 patties. Heat remaining oil in a nonstick skillet over medium heat. Add patties and cook, turning to brown both sides, about 5 minutes.

While patties are cooking, combine tomato, cucumber, cilantro and mint in a small bowl.

To serve, open pita bread halves and place a patty in each. Let each person top the patty with some of the diced tomato mixture.

Makes 4 servings.

Italian-Style Portobello Mushroom Sandwiches

Walking around our local agricultural fair, you will find one of the most seductive aromas is that of onions and peppers being cooked up to top sausage sandwiches. The same combination makes a delicious accompaniment to huge and tasty portobello mushroom caps, a fantastic vegetarian alternative!

1 tablespoon extra-virgin olive oil
1 large onion, thinly sliced
1 red bell pepper, cut into thin,
    2-inch strips
2 garlic cloves, minced

4 large portobello mushrooms
1 tablespoon tamari soy sauce
8 slices Italian bread
1 tablespoon minced fresh parsley

Heat oil in a large, nonstick skillet over medium heat. Add the onion and bell pepper and cook, stirring frequently, until vegetables are firm-tender, about 5 to 7 minutes. Add garlic and cook, stirring, 1 minute.

Remove stems from mushroom caps. (The stems can be saved to make vegetable broth for soups; see pages 12–14). Push the onion and bell pepper to one side of the skillet and place the mushroom caps, right side up, in the skillet. Drizzle soy sauce over the mushroom caps and reduce heat to low. Cook mushrooms 2 to 3 minutes. Turn, and cook mushrooms, covered, 2 to 3 minutes.

To serve, place a mushroom on each of 4 slices of bread. Top with some of the onion mixture, sprinkle with parsley and top with the remaining slices of bread.

Makes 4 servings.

Baba Pita

Baba ghannouj *is a wonderful, smoky-flavored dip of the Middle East. To duplicate the original flavor, grill the eggplant over a charcoal fire.*

1 medium eggplant (about 3/4
　　to 1 lb.)
1/4 cup sesame tahini
3 tablespoons fresh lemon juice
1 tablespoon extra-virgin olive oil
1/2 teaspoon salt
2 garlic cloves, minced

1 tablespoon fresh parsley
4 whole-wheat pita breads, halved
2 cups shredded cabbage
1 cup shredded carrots
1/2 cup thinly sliced celery
1/4 cup thinly sliced green onions

Preheat charcoal grill. Place the eggplant on the rack over medium heat 30 to 40 minutes, turning frequently, or until the eggplant is soft and the skin is lightly charred. (Alternately, the eggplant can be roasted in a 400F (205C) oven on a rack over a baking pan to catch drips). Trim off stem end and peel off skin while eggplant is hot.

Place eggplant flesh, tahini, lemon juice, oil, salt, garlic and parsley in a food processor and process until smooth.

To serve, open pita bread halves and spoon some of the eggplant mixture into each pita half. Top with a selection of cabbage, carrots, celery or green onions.

Makes 4 servings.

Black Bean Burgers with Salsa Guacamole

One of the great delights of these burgers is their wonderful aroma as they cook!

2/3 cup bulgur
1 1/4 cups boiling water
1 2/3 cup cooked (or 15-oz. can)
   black beans, drained
1/3 cup grated onion, squeezed dry
1 tablespoon extra-virgin olive oil
1 tablespoon fresh parsley
1 garlic clove, minced
1/2 jalapeño chile, seeded and
   coarsely chopped
1/4 teaspoon salt
1/4 teaspoon chili powder

1/4 teaspoon ground cumin
Dash of freshly ground black pepper
1/2 cup stone-ground yellow
   cornmeal
About 1 tablespoon vegetable oil
1 avocado, peeled, halved and pitted
1/4 cup salsa
1 teaspoon minced fresh cilantro
4 hamburger rolls
1 tomato, peeled, seeded and diced
2 green onions, thinly sliced on the
   diagonal

Place bulgur in a small bowl and cover with boiling water. Let stand 10 minutes. Drain well. Place soaked bulgur in a food processor with the beans, onion, olive oil, parsley, garlic, jalapeño, salt, chili powder, cumin and pepper. Process until smooth, scraping down the sides of the bowl as necessary.

Form bean mixture into 4 patties. Carefully dredge patties in the cornmeal. Heat oil in a nonstick skillet over medium heat. Add patties and cook, turning to brown both sides, about 5 minutes.

While bean patties are cooking, mash the avocado in a small bowl and stir in salsa and cilantro.

To serve, place bean patties on rolls, top with avocado mixture, diced tomato and green onions.

Makes 4 servings.

Side Dishes

For ease, try the quick-cooking grain varieties for side dishes. Couscous, bulgur, quick-cooking barley, instant brown rice, microwave polenta and risotto allow you to assemble a side dish in minutes. These grains also complement the vegetables, beans and dairy foods in the main dishes.

Another quick idea is to cook extra amounts of plain brown rice, then refrigerate or freeze for future use. Bring 2 cups brown rice, 4 cups water and 1/2 teaspoon salt to a boil. Reduce heat, cover and simmer 40 to 45 minutes, until the water is absorbed. Allow to cool, then pack into self-sealing storage bags according to the amount you will probably use per meal. When you're in a hurry, just reheat rice in the microwave or steam over boiling water.

Persian Rice

Nearly effortless and very flavorful, rice can be baked in your oven, Persian-style. If you choose to bake long-grain white rice in place of brown rice, reduce water to 1 cup and bake 30 to 35 minutes. For even more flavorful rice, cook with a good-quality vegetable broth.

1 cup long-grain brown rice, rinsed
    and drained
1 medium onion
4 whole cloves

1 1/2 cups water
1 1/2 tablespoons butter or soy
    margarine
1/2 teaspoon salt

Preheat oven to 325F (165C). Lightly oil a medium casserole dish with a lid.

Place the rice in prepared casserole dish. Stick the onion with the cloves and place in the center of the rice. In a small saucepan, bring water, butter and salt to a boil. Pour over rice and cover casserole dish. Bake 1 hour.

If rice is not dry and fluffy, remove onion and cloves, fluff rice with a fork and return to the oven 5 more minutes, uncovered.

Makes 4 servings.

Education

Although rice is thought of mainly as a side dish in the United States, in other parts of the world, rice is the very basis of most meals. To show the contrast, while here people eat an average of 17 pounds of rice per year (although the figure is growing), in Asia the average annual consumption is 300 pounds! Although the United States produces a meager 2 percent of the world's rice crop, our low consumption makes us a major exporter of the grain.

Savory Mixed-Grain Pilaf

A delicious combination of grains and vegetables with creative seasonings makes a versatile side dish. This is a very satisfying dish.

1 tablespoon olive oil
1 small onion, chopped
2 garlic cloves, minced
1/4 teaspoon *each* mustard seeds,
    ground cumin, ground
    coriander, paprika and salt
1/8 teaspoon freshly ground black
    pepper

1/4 cup minced carrot
1/4 cup minced celery
1/4 cup minced fresh mushrooms
1 cup converted long-grain rice
1/4 cup bulgur
2 cups vegetable broth
1 tablespoon fresh lemon juice

Heat the oil in a large saucepan over medium heat. Add onion and cook, stirring frequently, until the onion is translucent, about 2 to 3 minutes. Add the garlic, mustard seeds, cumin, coriander, paprika, salt and pepper and cook, stirring, 1 minute. Add carrot, celery and mushrooms and cook, stirring frequently, 2 to 3 minutes.

Add rice, bulgur, broth and lemon juice and bring to a boil. Stir, reduce heat, cover and simmer 20 minutes or until grains are tender and liquid is absorbed. Remove from heat and let stand 5 minutes before serving.

Makes 4 servings.

Education

Variety in your daily menu allows you to take advantage of a wide range of vitamins, minerals and trace elements in foods. One way to increase variety is with combinations of grains, as above.

Sweet Mixed-Grain Pilaf

A flavorful and exotic combination of grains, dried fruit and nuts provide a wide range of nutrients and pleasing flavor. This makes enough for company, or freeze for later what you don't use today.

1 tablespoon canola oil	1 1/2 cups long-grain brown rice
1 shallot, minced	1/2 cup wild rice
1/2 teaspoon salt	1/2 cup barley or millet
1 teaspoon ground coriander	1/4 cup golden raisins
1/2 teaspoon ground cinnamon	1/4 cup finely chopped pitted dates
1/4 teaspoon ground cardamom	1/4 cup slivered almonds
3 cups vegetable broth	1 teaspoon freshly grated
2 cups apple juice	orange peel

Heat the oil in a large saucepan over medium heat. Add the shallot and cook, stirring frequently, until the shallot is translucent, about 2 to 3 minutes. Add the salt, coriander, cinnamon and cardamom and stir until well combined.

Add the vegetable broth and apple juice and bring to a boil. Stir in brown rice, wild rice, barley or millet, raisins, dates, nuts and orange peel. Return to a boil. Reduce heat, cover and simmer 45 minutes, or until liquid is absorbed and grains are tender. Fluff with a fork to combine ingredients well before serving.

Makes 8 servings.

Orange Almond Brown Rice

Instant brown rice is definitely convenient, but lacks the deep, satisfyingly nutty flavor of regular brown rice. A few seasoning twists can make up the difference.

2 teaspoons olive oil
1 shallot, minced
2 tablespoons slivered almonds
2 cups instant brown rice
1 1/2 cups water

3/4 cup freshly squeezed
    orange juice
1/4 teaspoon freshly grated
    orange peel
1/4 teaspoon salt

Heat the oil in a medium pan over medium heat. Add the shallot, stirring often, until the shallot is translucent, about 1 to 2 minutes. Add the almonds and stir 1 minute. Stir in rice until combined.

Add water, orange juice, orange peel and salt and bring to a boil. Reduce heat, cover and simmer 10 minutes, or until the liquid is absorbed. Let stand 1 to 2 minutes, then stir and serve.

Makes 4 servings.

Education

Both nuts and rice complement beans in yielding optimal protein from a meal. This side dish contains both. In addition to the appealing flavor and texture of the added almonds, the rice gets a nutritional boost, too.

Portobello Pilaf

Especially nice with grilled foods! A squeeze of lemon juice brings out the flavor.

2 teaspoons olive oil
1/2 cup thinly sliced green onions
1/2 teaspoon ground coriander
Dash of freshly grated nutmeg
Dash of freshly ground black
    pepper
Dash of salt
2 cups chopped portobello
    mushroom caps

1/2 cup bulgur
1/2 cup quick-cooking barley
2 cups vegetable broth
1/2 cup diced red bell pepper
1 tablespoon minced fresh parsley
1/2 teaspoon minced fresh tarragon
Lemon wedges, for garnish

Heat the oil in a medium saucepan over medium heat. Add the green onions and cook, stirring frequently, until wilted, about 1 minute. Add ground coriander, nutmeg, black pepper and salt and stir until combined.

Add mushrooms and cook, stirring frequently, until mushrooms are tender, about 4 to 5 minutes. Stir in bulgur and barley. Add broth and bring to a boil. Stir in bell pepper, parsley and tarragon. Reduce heat, cover and simmer 10 minutes.

Remove from heat and let stand for 5 minutes, covered. Serve with lemon wedges.

Makes 4 servings.

Polenta

The perfect dish for the microwave. A versatile side dish, or the basis for a meal (see page 122), quick-and-easy polenta is sure to become a favorite!

3 cups water
2/3 cup stone-ground yellow
    cornmeal

1 teaspoon salt
1 tablespoon butter or soy
    margarine

Combine all ingredients in a large microwave-safe bowl. Microwave, uncovered, on HIGH 4 minutes or just until mixture begins to thicken. See Note, below. Stir well and microwave on HIGH 4 minutes. Remove from oven, stir, and cover bowl with a plate. Let stand 3 minutes before serving.

Makes 6 servings.

Note Cooking times will vary, depending on the wattage of your oven. Polenta should thicken and form soft lumps that can be stirred smooth. (Hard lumps indicate overcooking; use a food processor to blend out hard lumps and continue cooking.)

Lemon Pecan Couscous

A beautiful side dish! Instant couscous cooks in 5 minutes, which makes it ultra-convenient. Cumin and cinnamon add a subtle, exotic note with pecans and currants.

2 teaspoons canola oil
2 tablespoons minced carrot
2 tablespoons minced green onion
1/4 cup chopped pecans
1/4 teaspoon ground cumin
1/4 teaspoon ground cinnamon
1/4 cup dried currants or chopped
    raisins

1 1/2 cups couscous
1 1/2 cups apple juice
1 cup vegetable broth
2 tablespoons fresh lemon juice
1/4 teaspoon grated lemon peel

Heat the oil in a medium saucepan over medium heat. Add the carrot, green onion and pecans and cook, stirring frequently, until the pecans are lightly browned and fragrant, about 3 to 4 minutes. Stir in cumin and cinnamon until well combined.

Stir in currants or raisins and couscous. Add apple juice, broth, lemon juice and lemon peel and stir until combined. Bring to a boil over medium-high heat. Reduce heat to low, cover and steam 5 minutes.

Remove pan from heat and let stand 2 minutes. Fluff with a fork and serve.

Makes 6 servings.

Parslied Bulgur with Sun-Dried Tomatoes

Wonderful flavor and an attractive color. Sprinkle with a little freshly minced parsley when serving.

2 teaspoons olive oil
1/4 cup finely diced sun-dried
    tomatoes
1 small onion, chopped

1 cup bulgur
2 1/4 cups vegetable broth or water
1/2 teaspoon salt
1/4 cup minced fresh parsley

Heat the oil in a medium saucepan over medium-high heat. Add the tomatoes and onion and cook, stirring occasionally, until the onion is translucent, about 2 to 3 minutes.

Stir in bulgur. Add stock or water and salt and bring to a boil. Stir in parsley. Reduce heat, cover and simmer 15 minutes, or until liquid is absorbed and bulgur is soft. Let stand, covered, 5 minutes before serving.

Makes 4 servings.

Risotto with Garlic & Spinach

Again, the microwave is a perfect partner for this grain dish. Arborio rice, a must, is available in Italian markets and some supermarkets.

1 small onion, finely chopped
2 teaspoons olive oil
1 teaspoon soy margarine
1 large garlic clove, minced

3/4 cup Arborio rice
2 cups hot vegetable broth
1/2 cup lightly packed, chopped
    fresh spinach leaves

Combine the onion, oil and margarine in a large microwave-safe bowl. Microwave, uncovered, on HIGH 1 minute. Stir and microwave, uncovered, on HIGH 1 minute. Stir in garlic, then add rice, stirring until individual grains are coated with oil. Microwave, uncovered, on HIGH 1 minute.

Stir vegetable broth into rice and microwave, uncovered, on HIGH 6 minutes. Stir and microwave, uncovered, on HIGH 6 minutes. Stir in spinach, cover bowl with waxed paper and microwave on HIGH 1 minute. Remove from oven, stir, and cover bowl with a plate. Let stand 3 to 4 minutes before serving.

Makes 4 servings.

Rosy Rice

Use tomato juice to flavor instant brown rice, and to add an appealing color to grace your dinner plates.

1 1/2 cups water	2 teaspoons canola oil
2/3 cup tomato or eight-vegetable juice	1/4 teaspoon salt
2 cups instant brown rice	1 teaspoon minced fresh parsley, for garnish

Combine water and juice in a medium saucepan and bring to a boil over medium-high heat. Stir in remaining ingredients, except parsley, and return to a boil.

Reduce heat, cover and simmer 10 minutes, or until rice is tender and water is absorbed. Serve garnished with parsley.

Makes 4 servings.

Variations

Onion-Tomato Rice Sauté 1 medium onion, finely chopped, in the oil until firm-tender, then add liquid and bring to a boil. Proceed as above with rice and salt.

Stir 1 cup cooked green peas into hot rice.

Sally's Green Beans

So simple, but festive: green beans dressed up with stewed tomatoes—most of the flavor is built right in, without fuss.

1 pound fresh green beans
1/2 cup water
2 garlic cloves, minced
1/4 teaspoon salt
1/8 teaspoon freshly ground black
    pepper

1 (14-oz.) can Mexican or
    Italian-style stewed tomatoes,
    chopped, with liquid
1 tablespoon minced fresh parsley

Trim and cut the green beans in half. Bring water to a boil in a large saucepan over medium-high heat. Add beans and garlic and return to a boil. Reduce heat, cover and simmer 8 to 12 minutes, or just until firm-tender. Drain.

Stir in stewed tomatoes and parsley. Heat through and serve.

Makes 6 servings.

Garlic Mashed Potatoes

If it's hard to imagine mashed potatoes without lots of butter and whole milk, this lighter style will get you on a different track.

3 to 4 large, red potatoes, peeled
    and cubed
1/4 cup low-fat milk
3 garlic cloves, pushed through a
    press

2 teaspoons olive oil
1/4 teaspoon salt
Dash of freshly ground pepper
Dash of freshly grated nutmeg
1/4 cup low-fat yogurt

Place cubed potatoes in a large saucepan with enough water to generously cover and bring to a boil over medium-high heat. Reduce heat and simmer until very tender, about 15 minutes.

Meanwhile, combine the milk, garlic, oil, salt, pepper and nutmeg in a small saucepan. Bring to a boil over medium-high heat. Reduce heat and simmer 2 minutes. Turn off heat and let stand. When it has cooled slightly, stir in yogurt.

When potatoes are done, drain and return to pan. Mash, adding enough of the garlic mixture to obtain desired consistency.

Makes 4 servings.

Variation

Substitute vegetable broth for milk and yogurt for a dairy-free dish.

Sautéed Cherry Tomatoes with Fresh Herbs

A side dish that adds color and flavor contrast to any meal and can be made up in a couple of short minutes—what more can a cook ask? Blanching and peeling are optional.

1 (12-oz.) basket cherry tomatoes
1/2 tablespoon extra-virgin olive oil
1 teaspoon *each* minced fresh basil,
    fresh oregano or thyme, fresh
    tarragon or dill and fresh
    parsley

Dash of salt
Dash of freshly ground pepper

Bring a large saucepan of water to a boil. Add tomatoes and cook 30 seconds. Drain and rinse under cold water until cool. Peel tomatoes.

When ready to serve, heat oil in a large skillet. Add tomatoes and herbs and cook, stirring, 1 minute. Season with salt and pepper and place in serving bowl.

Makes 4 servings.

Red New Potatoes with Fresh Chives

Small red potatoes studded with fresh chives make a colorful side dish.

1 1/2 pounds small, red potatoes, scrubbed
3 cups vegetable broth
1 garlic clove, peeled

1 bay leaf
1/3 cup minced fresh chives
Dash of salt
Dash of freshly ground pepper

Combine the potatoes, broth, garlic and bay leaf in a large saucepan and bring to a boil. Reduce heat and simmer 15 minutes, or until potatoes are tender. Add chives, then drain potatoes in a strainer. Place potatoes and chives in a serving bowl and season with salt and pepper.

Makes 4 servings.

Roasted Herbed Vegetable Medley

An occasional stirring is all it takes to make this colorful dish.

2 pounds small, new potatoes, halved
1 *each* red, green and yellow bell peppers, cut into small pieces
1 red onion, cut into wedges
1 small fennel bulb, cut into wedges
3 tablespoons fresh lemon juice

3 tablespoons olive oil
6 garlic cloves, minced
1 teaspoon grated lemon peel
1/2 teaspoon dried rosemary, crumbled
1/2 teaspoon salt
1/4 cup chopped ripe olives

Preheat oven to 450F (230C). Combine the vegetables, lemon juice, oil, garlic, lemon peel, rosemary and salt in a 13 x 9-inch baking pan. Bake 55 to 60 minutes, stirring occasionally. Stir in the olives during the final 5 to 10 minutes of baking. Serve hot or at room temperature.

Makes 6 servings.

Easy Sweet Potato & Pineapple Casserole

Choose the big, orange sweet potatoes, sometimes called "yams." Assemble this side dish in minutes, then let it slowly bake. It makes an aromatic accompaniment to many meals.

2 large sweet potatoes, peeled
1/2 Granny Smith apple, peeled
   and cubed
1 (8-oz.) can pineapple chunks,
   drained and juice reserved

Dash of ground cinnamon
Dash of freshly grated nutmeg
Dash of salt
1/2 cup water

Preheat oven to 350F (175C). Lightly oil a 13 x 9-inch casserole dish.

Slice the sweet potatoes crosswise into 1/4-inch-thick slices. Toss in prepared casserole dish with the apple and pineapple chunks. Season with cinnamon, nutmeg and salt. Pour pineapple liquid and water over mixture and cover with foil. Bake 60 to 90 minutes, or until very tender.

Makes 6 servings.

Orange-Basil Baked Tomatoes

Another colorful "side" for the vegetarian dinner. Orange peel lends an unexpectedly refreshing flavor.

2 large, garden-ripe tomatoes,
    halved crosswise
1 garlic clove, minced
1/4 teaspoon finely grated
    orange peel

1/2 tablespoon butter or soy
    margarine, at room
    temperature
1 teaspoon basil leaves, thinly sliced

Preheat oven to 400F (205C).

Arrange tomatoes, cut side up, in a shallow casserole dish. Trim the blossom ends, if necessary, so they can stand flat. Stir the garlic and orange peel into the margarine.

Spread the orange mixture equally over each tomato half. Bake 25 minutes. Sprinkle with basil, cover with foil and bake 10 minutes, or until tomatoes are tender.

Makes 4 servings.

Creamy Cabbage

Cabbage is so healthful that you will want to find many ways to serve it. Coleslaw may be the most familiar cabbage dish, but this dish is so delicious, it will encourage even the most reluctant to eat more cabbage.

6 cups shredded cabbage
3/4 cup low-fat milk
1 tablespoon butter or soy margarine

Dash of salt
Dash of pepper

Place the shredded cabbage and milk in a large saucepan and bring to a boil over medium heat. Reduce heat, cover and simmer 10 to 15 minutes, or until the cabbage is tender. Toss with butter, salt and pepper and serve in a warmed bowl.

Makes 4 servings.

Education

If you don't already like cabbage, experiment with ways of preparing it so that you can happily include more of it in your diet. Cabbage contains compounds called "indoles" as well as large amounts of fiber, both of which appear to lower the risk of various forms of cancer.

Squash, Leek & Parsnip Puree

Vegetable purees are a terrific accompaniment to vegetarian main courses. Create color, flavor and aroma with apple cider and a sprinkling of spices.

2 cups cubed butternut squash
1 1/2 cups chopped leeks, white only
2 small parsnips, sliced crosswise
1/4 cup apple cider
2 teaspoons butter or soy margarine

1 teaspoon fresh lemon juice
1/2 teaspoon ground coriander
1/4 teaspoon ground cinnamon
1/8 teaspoon freshly grated nutmeg
1/8 teaspoon salt

Combine the squash, leeks and parsnips in a large saucepan with water to cover and bring to a boil over medium-high heat. Reduce heat, cover and simmer 12 to 15 minutes, or until vegetables are tender. Drain, reserving liquid for use as a vegetable stock, if desired.

Place vegetables in a food processor with remaining ingredients and process until smooth. Return to saucepan and heat through.

Makes 4 servings.

Education

Although at one time parsnips were quite popular, they were gradually replaced by potatoes on the majority of dinner tables. Parsnips, a good source of vitamin C, are also rich in fiber. Try them in the recipe above, or slice and cook with carrots. You'll enjoy their sweet, nutty flavor.

Farmhouse Home Fries

The microwave shortens the cooking time for these delicious home fries with old-time flavor, but little fat.

2 medium red potatoes (about 1/2 pound)	Dash of salt
	Dash of freshly ground black pepper
1 teaspoon canola or olive oil	Dash of paprika

Peel and thinly slice potatoes. Arrange on a luncheon-size microwave-safe plate, overlapping slices, but spreading them out over plate. Microwave, uncovered, on HIGH 2 minutes. Turn with a spatula. Microwave, uncovered, on HIGH 2 minutes.

Heat oil in a medium nonstick skillet over medium-low heat. Place potatoes in skillet with the spatula, keeping the slices in the same pattern. Sprinkle with salt, pepper and paprika, and cook 3 or 4 minutes, or until lightly browned on underside. Turn and brown remaining side. Cut in half.

Makes 2 servings.

Note To serve four people, thinly slice 3 large red potatoes. Arrange in an even layer on a dinner-size microwave-safe plate. Microwave, uncovered, on HIGH 5 minutes. Place another dinner plate over the potatoes and flip, transferring potatoes to the second plate. Microwave, uncovered, on HIGH 5 minutes. Transfer to a large, preheated, nonstick skillet with 2 teaspoons oil and/or butter. Cook on both sides until browned to perfection, seasoning as potatoes cook.

Breads & Breakfasts

Bread is called "the staff of life" and breakfast is "the most important meal of the day." Pretty heady stuff for such simple fare. But simplicity need never be boring.

This chapter gives you a special technique for increasing the flavor of yeast breads. (If you've never ventured to bake some, give it a try! It may provide you with a special satisfaction you've not experienced before . . .) Several quick breads show that even easy baking can yield mouth-watering flavors. Muffins, popovers and flat bread are other possibilities for you to try.

Breakfasts

Since so many breakfasts consist of cereal with milk, I've created some dairy-free alternatives. Fruit juice, nuts, dried fruits and grains give breakfast the nutritional re-spect it deserves if it is going to carry the "most important" mantle. While there is a blender breakfast that can go with you on those very busy mornings, I hope the need is only occasional. Pancakes and French toast with fresh fruit sauces can be breakfast fare on more relaxed mornings.

If you're trying to cut back on dairy at breakfast, try these approaches:

- Make oatmeal thinner than usual when cooking, then stir in a teaspoon of fruit jam.

- Use cider or apple juice in place of milk on porridge and breakfast cereals.

- Spread breads with jam or a small amount of nut butters in place of cream cheese or butter.

- Check packaged bread and other produce labels for milk products such as whey.

Breads

In many parts of America today, bakeries are springing up to meet the growing demand for "real" breads. Their cases are filled with ryes and pumpernickels, mixed whole-grain and crusty wheat Italian loafs. If you are lucky enough to have such a bakery in your area, you'll probably want to patronize it. Along with a soup or hearty salad, such a substantial bread can make a meal. And, of course, your friendly baker has taken all of the work out of it for you.

Hints for Making Yeast Bread

But, if you don't have a venturesome baker, or for some other reason want to bake a great loaf of bread on your own (say, to impress someone you're beginning to hold dear . . .), it pays to know how to create at least one good yeast bread. Let's face it, nothing else smells quite as homey in the kitchen, and we're not even counting the incomparable flavor!

Here's a hint on that incomparable flavor. While nearly every bread recipe you've ever seen may have said something like "Cover the bowl and let stand in a warm place until doubled in bulk," I'm going to tell you to do just the opposite. The fact of the matter is, the faster a bread rises, the less flavor it has time to develop. So if you're going to the trouble of making a yeast loaf at all, you may want to make sure it's as tasty as it can be!

First, mix up the ingredients of the following recipe the night before you want to bake bread. Place the dough in a large bowl and cover tightly with plastic wrap to keep the dough from drying out. (Use a large bowl because this little ball of dough is going to expand. And you don't want to give yourself another job of cleaning up an unnecessary mess.)

Now, set the bowl in the refrigerator and ignore the dough until the next morning. That's right! The yeast, which is a tiny living organism, may shiver, but it will still do its job to raise the dough properly. The yeast will do its work untended, and you can read a book, get a good night's sleep, or go out on the town.

Next morning, take out the dough and punch out all the air. Form the dough into a long loaf (special bread pans are available, or you can fashion something out of foil, oiled and rolled up at the sides to keep the dough from spreading out) or roll up and then pat the dough into the bottom of a lightly oiled, nonstick 9 x 5-inch loaf pan. Cover with lightly oiled waxed paper so that the top doesn't dry out, then let the loaf stand on your counter until it has warmed to room temperature and doubled in bulk.

Generally, I find the bread is ready to bake just in time to be ready for dinner. Again, there's no hurry here. This could take several hours. But on the other hand, since the process is a slow one, the bread doesn't need a lot of baby-sitting.

One Terrific Loaf of Bread

Don't forget to try some of the variations as you become more familiar with baking bread, or devise some signature loaves on your own, depending on the types of flour you have available.

1/4 cup lukewarm (110F; 45C) water
1 tablespoon honey
1 package active dry yeast (1 scant tablespoon)
1 1/4 cups hot water
1 1/2 teaspoons salt

2 cups unbleached all-purpose or bread flour
3/4 cup plain wheat germ
1/4 cup olive oil
About 1 1/2 cups whole-wheat flour or whole-wheat pastry flour

Stir together the lukewarm water and honey in a warmed small bowl (to keep the water from cooling too quickly). Sprinkle on the yeast and set aside until mixture is frothy, about 5 minutes. Combine the hot water and salt in a large bowl, and let stand until salt is dissolved, and water has cooled to lukewarm.

Add the yeast mixture to the large bowl, along with the all-purpose or bread flour. Beat with a wooden spoon until the dough appears "stretchy." Add the wheat germ and olive oil and enough of the whole-wheat flour to make a firm, but pliable, dough. (If you lightly massage your cheeks, that's about what the bread dough should feel like. Try it before your hands are covered in flour . . .)

Now, either follow the directions given under Hints for Making Yeast Bread (page 194), or place the bowl in a pan of warm water and cover with lightly oiled plastic wrap. Let the dough rise until doubled, 1 to 1 1/2 hours.

Punch down the dough and form into a loaf shape. Pat into the bottom of a lightly oiled 9 x 5-inch nonstick bread pan. Cover with lightly oiled waxed paper, so that the top doesn't dry out. Let rise again until the dough reaches the top of the pan, or is about doubled, 45 to 60 minutes.

Preheat the oven to 425F (220C). Bake the bread 10 minutes, then turn the heat to 325F (165C) and bake about 30 to 40 minutes more, or until browned and crusty.

Makes 1 loaf.

Variations

Decrease the wheat germ to 1/2 cup and add 1/4 cup stone-ground yellow cornmeal.

Omit the wheat germ and add 3/4 cup rye flour. Combine rye, cornmeal and wheat germ to equal 3/4 cup total.

For any of these variations, you can substitute molasses for the honey.

Scalded buttermilk can be substituted for the hot water to dissolve salt.

To give the bread a professional finish, dust with some of the unbleached flour before baking.

Cape Cod Carrot Bread

A not-too-sweet whole-grain bread with a wonderful texture and delicate flavor.

1 1/4 cups whole-wheat flour	1/2 teaspoon salt
3/4 cup plain wheat germ	2 eggs
1/3 cup stone-ground yellow cornmeal	1/2 cup fresh orange juice
1/2 cup packed light brown sugar	4 tablespoons soy margarine or butter, melted
2 teaspoons baking powder	2 tablespoons maple syrup
1 teaspoon baking soda	2 cups shredded carrots
1 teaspoon ground cinnamon	

Preheat oven to 400F (205C). Lightly oil a 9 x 5-inch nonstick loaf pan.

Combine flour, wheat germ, cornmeal, brown sugar, baking powder, soda, cinnamon and salt in a large bowl. In a medium bowl, lightly beat the eggs and stir in orange juice, margarine and maple syrup. Add with carrots to the dry ingredients and stir just until combined.

Spoon the batter into prepared pan and place in the oven. Immediately reduce heat to 350F (175C) and bake 60 to 70 minutes, until a wooden pick inserted in the center comes out clean. Cool in pan 10 minutes. Turn out on a wire rack to finish cooling. Cut into slices.

Makes 1 loaf.

Golden Apricot Tea Bread

You'll love every nibble of this delicately flavored, attractive bread. Served with herbal tea, this bread could single-handedly initiate a tradition of afternoon tea in your home!

1 cup dried apricots, finely diced
1/4 cup fresh cranberries, finely chopped
1/3 cup fresh orange juice
2 tablespoons butter or soy margarine
1/2 cup packed light brown sugar
2 cups whole-wheat flour

1 cup unbleached all-purpose flour
2 teaspoons baking powder
1 teaspoon baking soda
1 teaspoon salt
1 egg
1 1/2 cups buttermilk
1 teaspoon granulated sugar

Preheat oven to 350F (175C). Lightly oil a 9 x 5-inch nonstick loaf pan.

Place the apricots, cranberries, and orange juice in a medium saucepan over low heat. Simmer, covered, 5 minutes. Remove from heat and stir in the butter until melted, then the sugar, until well blended.

Combine the flours, baking powder, baking soda and salt in a large bowl. Beat the egg in a medium bowl and stir in the buttermilk.

Combine the apricot mixture and buttermilk mixture with the dry ingredients, stirring just until combined. Spoon the batter into prepared pan. Bake 50 to 55 minutes, or until a wooden pick inserted in the center comes out clean. Cool in pan 10 minutes. Turn out on a wire rack to finish cooling. Cut into slices.

Makes 1 loaf.

Education

Dried apricots are a concentrated source of valuable nutrients: beta carotene, iron, niacin, potassium and even some calcium. Drying the apricots, however, reduces their vitamin C content. And they are high in sugar. Because they tend to stick to your teeth, brush or rinse your mouth after eating dried apricots to prevent cavities.

Indian Chapati Flat Bread

Simplicity itself, flat bread requires no yeast and is cooked on top of the stove. Serve with curried main dishes, soups and stews.

2 cups wheat-blend bread flour   1/2 teaspoon salt
    (see Note, below)   About 1/2 cup water

Combine the flour and salt in a large bowl. Add the water slowly, kneading, until dough is soft, but not sticky. (If it gets sticky, add more flour.) Knead dough on a floured surface 5 minutes. Or mix ingredients in a food processor, adding just enough water until dough holds together, and process 1 minute to knead. Place dough in a lightly oiled bowl. Cover with a towel and let dough rest 30 to 60 minutes.

Divide dough into 12 equal pieces and roll into balls. Roll out each ball as thin as possible on a floured surface.

Heat a dry skillet or griddle over medium heat until drops of water dance when sprinkled on the surface. Cook flat breads, turning once, until puffed and cooked through, about 2 minutes. Turn down heat if the breads brown too quickly. Keep warm until serving.

Makes 12 chapati.

Note If your supermarket does not have wheat-blend bread flour, use 1 1/3 cups unbleached bread flour and 2/3 cup whole-wheat flour instead.

Cornbread

A great accompaniment to bean dishes, especially highly seasoned Southwestern meals.

1 1/4 cups stone-ground yellow
    cornmeal
3/4 cup unbleached flour
1 teaspoon baking powder
1/4 teaspoon salt

1 egg
1 to 1 1/4 cups buttermilk
3 tablespoons unsalted butter,
    melted
1 tablespoon light brown sugar

Preheat oven to 425F (220C). Lightly oil an 8-inch round nonstick baking pan.

Mix cornmeal, flour, baking powder and salt in a medium bowl. Beat egg in a large bowl. Stir in 1 cup of the buttermilk, melted butter and brown sugar. Add dry ingredients and stir just until combined, adding extra buttermilk if the batter seems dry.

Spoon batter into prepared pan. Bake 25 minutes, or until a wooden pick inserted in the center comes out clean. Serve warm from the oven.

Makes 8 servings.

Education

Population studies in the United States have shown that vegetarians tend to be healthier and live longer than other Americans. Reduced fat, high fiber and a concentration on vegetables and grains, common among vegetarians, are all associated with a lowered risk of disease.

Quick Wheat Bread with Dried Fruits

Grated lemon peel adds zest to a flavorful loaf. Vary the dried fruits, if desired.

3 cups whole-wheat pastry flour
1 tablespoon baking powder
1/4 teaspoon salt
1/2 cup raisins
1/4 cup chopped, pitted dates
1/4 cup chopped, pitted prunes

3 tablespoons maple syrup
2 tablespoons canola oil
1 tablespoon sesame tahini
1 1/2 cups apple juice
Grated peel of 1 lemon

Preheat oven to 350F (175C). Lightly oil a 9 x 5-inch nonstick loaf pan.

Combine flour, baking powder and salt in a large bowl. Stir in dried fruits. In a small bowl, combine maple syrup, oil and tahini. Slowly stir in 1/2 cup of the juice, and the lemon peel. Add to dry ingredients with remaining 1 cup juice and stir just until combined.

Spoon batter into prepared pan. Bake 1 hour, or until a wooden pick inserted in the center comes out clean. Cool in pan 10 minutes. Turn out on a wire rack to finish cooling. Cut into slices.

Makes 1 loaf.

Orange-Raisin Carrot Muffins

Delicious warm from the oven with a little marmalade.

1/2 cup packed light brown sugar
2 tablespoons soy margarine or
    butter, softened
1 egg, beaten
1/2 cup grated carrots
1/2 cup raisins
3/4 cup orange juice

1/2 teaspoon freshly grated
    orange peel
2 cups unbleached all-purpose flour
2 teaspoons baking powder
1 teaspoon ground cinnamon
1/4 teaspoon salt

Preheat oven to 350F (175C). Lightly oil 12 nonstick muffin cups.

Beat together the brown sugar and margarine or butter until creamy. Stir in egg until well combined. Stir in carrots, raisins, orange juice and orange peel.

Combine flour, baking powder, cinnamon and salt in a small bowl. Stir into carrot mixture just until combined.

Spoon batter into prepared muffin cups. Bake 25 to 30 minutes, or until a wooden pick inserted in the center of a muffin comes out clean. Turn muffins out on a wire rack to cool.

Makes 12 muffins.

Gingerbread Date-Nut Muffins

Serve topped with applesauce for a delicious breakfast treat.

1/3 cup packed light brown sugar
2 tablespoons soy margarine or
　　butter, softened
1/3 cup chopped pitted dates
1/4 cup chopped walnuts
2 tablespoons medium unsulfured
　　molasses
1 egg, beaten
1 cup whole-wheat flour

1 cup unbleached all-purpose flour
1 1/2 teaspoons baking powder
1 teaspoon ground cinnamon
1 teaspoon ground ginger
1/4 teaspoon ground cloves
1/4 teaspoon salt
1 cup apple juice, cider or pineapple
　　juice

Preheat oven to 350F (175C). Lightly oil 12 nonstick muffin cups.

Beat together the brown sugar and margarine until creamy. Stir in chopped dates, walnuts, molasses and egg until well combined.

Combine flours, baking powder, cinnamon, ginger, cloves and salt in a small bowl. Add with the juice to the date mixture and stir just until combined.

Spoon batter into prepared muffin cups. Bake 25 to 30 minutes, or until a wooden pick inserted in the center of a muffin comes out clean. Turn muffins out on a wire rack to cool.

Makes 12 muffins.

Education

Applesauce is a terrific toast or muffin topper in the morning (or for a late-night snack). It is moist and has no fat. Sweet, yet low in sugar, applesauce also contains fiber. It's a real flavor and nutrition plus.

Nantucket Blueberry Muffins

Easy to describe in one word: scrumptious!

1 cup whole-wheat flour	1 cup fresh or frozen (unthawed)
1 cup unbleached all-purpose flour	blueberries
1/2 cup plain wheat germ	2 eggs
2 teaspoons baking powder	3/4 cup buttermilk
1/2 teaspoon baking soda	1/3 cup maple syrup
1 cup chopped walnuts	2 tablespoons canola oil

Preheat oven to 400F (205C). Lightly oil 12 nonstick muffin cups.

Combine the flours, wheat germ, baking powder and soda in a large bowl. Stir in the walnuts and blueberries.

Beat eggs in a medium bowl. Stir in buttermilk, maple syrup and oil. Stir the egg mixture into the dry ingredients just until combined.

Spoon batter into prepared muffin cups. Bake 20 to 25 minutes, or until a wooden pick inserted in the center of a muffin comes out clean. Turn muffins out on a wire rack to cool.

Makes 12 muffins.

Education

Wheat germ is rich in polyunsaturated fat, which is one of the reasons it is commonly removed from flour. Because fats are inclined to go rancid, removing the germ from flour greatly extends the flour's shelf life. (When the germ and bran are removed from whole-wheat flour, the resulting flour is white.) Because wheat germ contains fat, it should be kept refrigerated.

Whole-Wheat Popovers

Popovers can be one of the delights of a summer's fresh fruit breakfast. Popovers are also wonderful for dinner. In winter, serve with all-fruit jams. Happily, there's no yeast involved, so the mixing is fairly easy. And, considering the results, you'll probably agree they are well worth the small expenditure of effort!

1/2 cup whole-wheat flour	1/4 teaspoon salt
1/2 cup unbleached all-purpose flour	2 eggs
1 cup low-fat milk	1 additional egg white
1 tablespoon canola oil	Additional whole-wheat flour
1 teaspoon packed light brown sugar	

Preheat oven to 450F (230C). Lightly oil 12 muffin cups and dust with whole-wheat flour.

Combine the flours, milk, oil, sugar and salt in a large bowl. Beat the eggs and egg white in a medium bowl with a fork or an electric beater until frothy. Add to the flour mixture and beat until well combined.

Fill each cup about two-thirds full with batter. Bake 15 minutes. Reduce heat to 350F (175C). Bake 20 to 25 minutes, or until popovers are crusty outside and moist inside. Remove from oven, slit each popover and open slightly to allow steam to escape. Serve at once.

Makes 12 popovers.

Variation

Cheese Popovers Fill each muffin cup one-third full of batter. Sprinkle with 1 to 2 teaspoons grated Cheddar cheese. Cover with remaining batter. Bake as above.

Quick Pecan Cinnamon Rolls

Biscuit dough gets a glamorous treatment for breakfast. These could easily become a Sunday morning ritual.

1 cup whole-wheat flour
1 cup unbleached all-purpose flour
1 tablespoon baking powder
1/2 teaspoon salt
4 tablespoons soy margarine or
    butter, chilled
1 egg, beaten

1/3 to 1/2 cup low-fat milk
2 tablespoons soy margarine or
    butter, at room temperature
1/4 cup pecans
1/4 cup packed light brown sugar
2 teaspoons ground cinnamon

Preheat oven to 375F (190C). Lightly oil a 9-inch-square nonstick baking pan.

Combine the flours, baking powder and salt in a large bowl. Cut in butter with a pastry blender until the mixture resembles crumbs. Combine egg with 1/3 cup milk and stir into dry ingredients just until combined, adding a little extra milk, if needed, to make a soft dough that holds together. (If mixture gets too wet, just add a little flour.)

On a floured surface, roll out the dough into a 14 x 6-inch rectangle. Spread with margarine or butter. Combine the pecans, sugar and cinnamon in a food processor and process until fine. Sprinkle evenly over the margarine. Beginning at a long side, roll up into a log. Cut crosswise into slices about 3/4 inch thick and arrange, cut sides down, close together in prepared baking pan.

Bake 15 to 18 minutes or until well browned. Let cool slightly, then separate with a table knife.

Makes 15 or 16 rolls.

Variations

Orange Marmalade Rolls Spread dough with half the amount of soft margarine or butter and spread with orange marmalade. Roll up, cut and bake as above.

Jam Rolls Use strawberry or other favorite jam in place of marmalade.

Raspberry Applesauce Coffeecake

Dots of raspberry jam add a cheerful note to this breakfast cake.

1/3 cup packed light brown sugar
2 tablespoons soy margarine or
 butter, softened
1 egg, beaten
1/2 cup applesauce
1/3 cup buttermilk
1/2 teaspoon vanilla extract
3/4 cup unbleached all-purpose
 flour

1/4 cup whole-wheat flour
1 teaspoon baking powder
1/4 teaspoon salt
1 to 2 tablespoons seedless red
 raspberry jam
1/4 cup almonds
1 teaspoon sugar

Preheat oven to 350F (175C). Lightly oil an 8-inch round nonstick baking pan.

Beat together the brown sugar and margarine or butter in a medium-size bowl until creamy. Stir in the egg, applesauce, buttermilk and vanilla.

Combine the flours, baking powder and salt in a small bowl. Stir into applesauce mixture just until combined. Spread batter in prepared baking pan. Dot the surface with 1/4 teaspoonfuls of jam.

Place the almonds and sugar in a food processor and process until finely ground. Sprinkle over the coffeecake. Bake 25 to 30 minutes, or until a wooden pick inserted in the center comes out clean.

Makes 6 servings.

Education

If you like baked goods, but don't want to wind up wearing them around your waist, bake with applesauce. Applesauce can impart a pleasing moist consistency to baked cakes and breads, cutting back on the need for fats, which also create a moist crumb. Because fats are twice as high in calories as an equal amount of carbohydrates (such as applesauce), this means a leaner final product.

Black Raspberry Breakfast Cake

A crumb topping and juicy black raspberries highlight an attractive coffee cake.

3/4 cup unbleached all-purpose
    flour
1/4 cup whole-wheat flour
1/3 cup packed light brown sugar
1/4 teaspoon salt
4 tablespoons soy margarine or
    butter, chilled
3 tablespoons finely chopped
    almonds, walnuts or pecans

1/8 teaspoon ground cinnamon
Dash of ground cardamom
Dash of freshly grated nutmeg
1/2 teaspoon baking powder
1 egg
1 additional egg white
1/4 cup buttermilk
1/2 cup fresh or frozen (unthawed)
    black raspberries

Preheat oven to 350F (175C). Lightly oil a 9 x 5-inch nonstick loaf pan.

Combine the flours, sugar and salt in a medium bowl. Cut in the margarine or butter with a pastry blender until the mixture resembles fine crumbs.

Remove 1/4 cup of the flour mixture and place in a small bowl. Add the nuts, cinnamon, cardamom and nutmeg. Set aside for topping.

To the remaining flour mixture, add baking powder and stir until combined. Beat egg and egg white lightly and add with buttermilk. Stir just until ingredients are combined.

Spread the batter in prepared baking pan. Top with raspberries and sprinkle with cinnamon crumb topping. Bake 30 to 35 minutes, or until a wooden pick inserted in the center comes out clean. Serve warm.

Makes 6 servings.

Variation

If black raspberries are not available, use red raspberries or blackberries.

Crushed Pecan French Toast

Baked, not fried, with a banana batter and crushed pecans, this French toast goes far beyond the ordinary. Top with fresh fruit and a light dusting of powdered sugar.

1/4 cup pecan halves	1/3 cup milk
1 tablespoon light brown sugar	1/4 cup mashed banana
1/4 teaspoon ground cinnamon	1/2 teaspoon vanilla extract
1 egg	8 slices raisin or French bread

Preheat oven to 450F (230C). Lightly oil a nonstick baking sheet.

Process the pecans, brown sugar and cinnamon in a food processor until crumbly. Remove and set aside in a small bowl.

Combine the egg, milk, banana and vanilla in the food processor and process until smooth. Place batter in a shallow bowl, large enough to accommodate the bread slices one at a time. Dip each slice into the batter, turning to coat both sides, and arrange on prepared baking sheet. Drizzle any remaining drops of batter over the slices. Sprinkle with half of the pecan mixture.

Bake 6 minutes, or until undersides of slices are browned. Turn and sprinkle with remaining pecan mixture. Bake 4 to 5 minutes, until remaining sides are browned and slices are baked throughout.

Makes 4 servings.

Berkshire Mountain Blueberry Pancakes

A non-egg, non-dairy pancake with great flavor. Top with added blueberries and a little maple syrup—delicious!

3/4 cup whole-wheat flour
1/4 cup unbleached all-purpose
    flour
1 teaspoon baking powder
Dash of ground cinnamon
2/3 cup apple cider or juice

2 tablespoons canola oil
1 tablespoon molasses
1/3 cup fresh or frozen (unthawed)
    blueberries
Additional blueberries, for garnish

Combine the flours, baking powder and cinnamon in a medium bowl. Combine cider or juice, oil and molasses in a small bowl. Stir into dry ingredients just until combined.

Heat a large nonstick skillet or griddle over medium heat. Lightly oil. Drop 1/4 of the batter onto griddle at a time, making 4 pancakes. Sprinkle each with some of the blueberries before tops are set. When undersides are golden brown, turn carefully and cook until browned and cooked through.

Makes 2 servings.

Sedona Red Rock Ricotta Pancakes

One of the highlights of a trip out West was an outdoor breakfast in red rock country that featured light and airy ricotta pancakes—wonderful!

3/4 cup whole-wheat flour
1/4 cup unbleached all-purpose
    flour
1 teaspoon sugar
1 teaspoon baking powder
1/2 teaspoon baking soda

1/4 teaspoon salt
1 egg
1/3 cup part-skim ricotta cheese
1 cup buttermilk
1 tablespoon canola oil

Combine the flours, sugar, baking powder, soda and salt in a large bowl. Beat the egg with the ricotta in a medium bowl. Stir in buttermilk and oil. Add ricotta mixture to the dry ingredients, stirring just until combined. (Add a little buttermilk or flour, if necessary, to obtain the batter consistency you prefer.)

Heat a large nonstick skillet or griddle over medium heat. Lightly oil. Drop batter onto griddle, making 4 smaller or 2 large pancakes. When undersides are golden brown, turn carefully and cook until browned and cooked through.

Makes 2 servings.

Strawberry Sauce

Use for pancakes or desserts. Add cornstarch if you want to serve the sauce warm.

3 cups sliced fresh strawberries
1/3 cup maple syrup

1 1/2 teaspoons cornstarch
(optional)

To serve cold, combine 2 cups of the strawberries and maple syrup in a blender and process until smooth. Stir in remaining 1 cup sliced strawberries.

To serve warm, add cornstarch to blender with 2 cups of the strawberries and maple syrup and process until smooth. Place blended mixture in a medium saucepan over medium heat and bring to a boil, stirring constantly. Cook, stirring constantly, until slightly thickened. Remove from heat and place in a serving bowl. Let cool slightly, then stir in remaining strawberries.

Makes 4 servings.

Molasses Pancakes

With buckwheat, cornmeal and whole-wheat flour, these pancakes offer a full morning's worth of high-powered energy!

1 cup whole-wheat flour
2 tablespoons buckwheat flour
2 tablespoons stone-ground yellow
    cornmeal
1 teaspoon baking soda

1 egg
1 tablespoon molasses
1 tablespoon canola oil
1 cup milk

Combine the flours, cornmeal and baking soda in a large bowl. Beat the egg lightly with molasses and oil in a small bowl. Stir in milk until combined. Add milk mixture to the dry ingredients, stirring just until combined.

Heat a large nonstick skillet or griddle over medium heat. Lightly oil. Drop batter onto griddle, about 1/4 cup for each pancake. When undersides are golden brown, turn carefully and cook until browned and cooked through.

Makes 3 to 4 servings.

Education

If you're sweet on molasses, you'll get a nutritional edge over white sugar. Molasses, a by-product of sugar refining, contains the vitamins and minerals. So, in addition to enjoying a pleasant sweetener with its own distinct and homey flavor, you'll also be getting thiamin, riboflavin, iron and calcium—vitamins and minerals your body needs to stay healthy.

Multigrain Buttermilk Pancakes

Use your own mix to make whipping up a pancake batter an easy matter in the morning.

1 1/2 cups Multigrain Pancake Mix (page 214)

1 egg

1 tablespoon canola oil

1 1/4 to 1 1/2 cups buttermilk

Place mix in a large bowl. Beat egg lightly with the oil in a small bowl. Add egg mixture and enough buttermilk to the mix to make a creamy batter.

Cook pancakes on a lightly oiled, nonstick griddle or skillet, turning when bubbles appear on the surface.

Makes 4 servings.

Variations

Use low-fat milk instead of buttermilk.

For a vegan pancake, omit egg and milk and use 2 tablespoons oil and 1 1/4 to 1 1/2 cups apple juice or other liquid.

Education

Whole grains contain both the bran layers and the germ, which is rich in oil. For this reason, keep this mix and your whole-grain flours in the refrigerator to keep them from turning rancid. For whole-grain flours that will not be used up within a few weeks, place the flour, in its original paper container, in a self-sealing freezer bag and freeze.

Multigrain Pancake Mix

Having a mix on hand makes pancake breakfasts an easy task, even on the most laid-back of weekends. To use, see recipe for Multigrain Buttermilk Pancakes (page 213).

3 cups unbleached all-purpose flour
2 cups quick-cooking rolled oats
1 1/2 cups whole-wheat flour
1/2 cup buckwheat flour
1/4 cup stone-ground yellow
    cornmeal

2 tablespoons sugar
1 1/2 tablespoons baking powder
1 teaspoon salt

Place the ingredients in a large self-sealing plastic bag. Seal and mix by turning bag back and forth. Keep refrigerated to preserve the freshness of the ground whole-grain flours.

Makes about 6 cups mix.

Thick Blueberry Breakfast Sauce

Serve over pancakes, French toast, or waffles. Frozen blueberries make this sauce a year-round enjoyment for breakfast, or even over desserts.

3 cups (1 pound) frozen blueberries
1 tablespoon cornstarch

1/3 cup maple syrup

Place frozen blueberries in a large microwave-safe bowl. Sprinkle with the cornstarch and toss until the blueberries are coated. Add maple syrup and stir until combined.

Microwave, uncovered, on HIGH 8 to 9 minutes, stirring twice, or until bubbly and thickened. Serve warm.

Makes 4 servings.

Cinnamon-Citrus Stewed Prunes

There is something comfortingly old-fashioned about stewed prunes. But if your diet features too little in the way of fiber, it can be a practical enjoyment as well.

1 pound large, pitted prunes
1 1/2 cups water
2 tablespoons fresh lime or lemon
    juice
2 tablespoons packed light brown
    sugar

1 cinnamon stick or dash of ground
    cinnamon
1/2 lemon, thinly sliced
Whipped cream, for garnish

Place the prunes in a small saucepan and add water, juice and brown sugar. Add cinnamon stick or dash of ground cinnamon. Layer the lemon slices over the prunes and bring to a boil. Reduce heat and simmer, uncovered, 30 minutes. Let cool.

Serve prunes in goblets with juice spooned over each serving, a slice of the lemon and a small dab of whipped cream for garnish.

Makes 4 servings.

Note If you use prunes with pits, remove pits before serving.

Education

This old-fashioned dish is updated with citrus and makes a beautiful presentation. But prunes are more than pretty: They contain more fiber than most other fruits and vegetables, including dried beans. In addition, over half of this is soluble fiber, the type that's been shown to lower blood cholesterol levels. Prunes are also rich in beta carotene, and contain valuable B vitamins, iron and potassium, another blood pressure "good guy."

Chocolate Blender Breakfast Bash

High in protein, nutritious, portable and tasty—perfect for those times when you want something smooth and simple for breakfast.

1 teaspoon unsweetened cocoa
   powder
1 teaspoon maple syrup
1 large ripe banana, broken into
   pieces

1 tablespoon peanut butter
1 cup low-fat milk

Combine the cocoa powder and maple syrup in a small dish until combined. Place in a blender with the banana and peanut butter and enough of the milk to make a smooth consistency as you blend the mixture. Add remaining milk and continue to blend until combined, scraping down the sides of the container as needed.

Makes 1 serving.

Variation

Banana Bash Omit cocoa.

Macadamia Nut Cider Oatmeal

This hearty and healthy dish is especially nice for autumn breakfasts.

1 cup water
1/2 cup quick-cooking rolled oats
1 tablespoon raisins
Dash salt

1/8 teaspoon ground cinnamon
2 macadamia nuts, chopped or sliced
1/4 cup apple cider

Combine water, oats, raisins, salt and cinnamon in a large microwave-safe bowl. Microwave, uncovered, on HIGH 1 1/2 to 2 minutes, or until creamy. Stir in nuts and cider.

Makes 1 serving.

Variations

For cooking on the stove, combine first four ingredients in a small saucepan and bring to a boil over medium heat. Cook, stirring occasionally, about 1 to 2 minutes. Continue as above.

Add 1/4 cup cubed apple before cooking. Substitute other nuts for macadamias.

Education

Most hot cereals are low in fat, or even fat free, and all are a valuable source of complex carbohydrates. To make the most of these healthy day-starters, keep a few things in mind:

- While most hot breakfast cereals are a bargain, at less than 15 cents per serving, instant varieties can cost two to three times as much.

- "Instant" hot cereal may not even save much time, since hot cereals can be microwaved right in your bowl!

- While regular and quick-cooking hot cereals are usually sodium- and sugar-free, instant varieties usually have salt added and many have added sugar.

- Oatmeal and other whole-grain cereals (such as Wheatena) have a good supply of B vitamins and important trace minerals found in the whole grain. During processing, refined cereals (such as grits and farina) lose fiber and many vitamins and minerals, of which only some may be replaced.

Sweet Apple Muesli

A simple breakfast, but substantial and satisfying.

1 Granny Smith apple, peeled,
   cored and diced
1/3 cup quick-cooking rolled oats
2 tablespoons ground almonds

2 tablespoons raisins
Dash of cinnamon
1/3 cup low-fat plain yogurt
   (optional)

Combine the diced apple in a serving bowl with the oats, ground almonds, raisins and cinnamon. Stir in yogurt, if desired.

Makes 1 serving.

Variations

Substitute 1 large pear for the apple. Use diced, pitted prunes or dates in place of raisins. Substitute chopped walnuts for almonds.

For a hot breakfast, add 2/3 cup water or cider and microwave, uncovered, on HIGH 2 to 2 1/2 minutes.

Education

Does an apple a day keep the doctor away? It definitely could help! Just one apple provides nearly a fifth of the amount of fiber nutrition experts say we should consume daily. Apples are a good source of both soluble and insoluble fiber. Studies show that soluble fiber is effective in lowering blood cholesterol levels; insoluble fiber is important to a healthy digestive system.

Apple-Raisin Breakfast Bowl

Bulgur, a precooked grain, is quick-cooking enough to be a breakfast option. If your supermarket does not carry plain bulgur (it is the main ingredient of tabbouleh), a Middle Eastern food store would have it.

1/4 cup bulgur
1 tablespoon chopped pecans
1 tablespoon raisins
2/3 cup boiling water
1/2 apple, peeled, cored and
    chopped

1 teaspoon honey or maple syrup
1/4 teaspoon ground cinnamon
Dash of freshly grated nutmeg
Dash of salt

Combine the bulgur, pecans, raisins and boiling water in a cereal bowl. Cover and set aside 5 minutes. Stir in apple, honey, cinnamon, nutmeg and salt. Microwave, uncovered, on HIGH 2 minutes. Stir well. Microwave, uncovered, on HIGH 1 minute, or until grain is firm-tender.

Makes 1 serving.

Education

Bulgur is a nutritious "convenience" grain. Minimally processed, bulgur is the result of steaming and drying whole-wheat kernels, which are then cracked into small pieces. Because it is precooked, it can be ready for eating in just a few minutes.

Sunny Morning Cereal Bowl

In winter, use apples, pears and bananas.

1/2 cup fresh or frozen blueberries
1/2 cup sliced strawberries
1/4 cup halved seedless grapes
1/2 cup rolled oats

1 tablespoon plain wheat germ
1 tablespoon sunflower kernels
1/4 cup orange juice (optional)

Combine the fruits in a cereal bowl and sprinkle with oats, wheat germ and sunflower kernels. Add orange juice if a moist cereal is desired.

Makes 1 serving.

Education

Sunflower seeds are rich in vitamin E: Just 1 tablespoon provides 20 percent of a woman's recommended daily allowance (about 17 percent of a man's). Sunflower seeds are also high in phosphorus, magnesium and fiber. Because they are high in fat, which makes them prone to rancidity, keep them in the refrigerator. Oh, and eat them in moderation for the same (high-fat) reason.

New England Porridge

An old-fashioned breakfast with cold-weather appeal. Try nuts and other fruits for variety. Adding diced bananas makes a creamy porridge without milk.

1/2 cup yellow stone-ground cornmeal
2 1/2 cups water
1/2 teaspoon salt

1/4 teaspoon ground cinnamon
2 large bananas, diced
3 tablespoons maple syrup

Stir together the cornmeal, water, salt and cinnamon in a large, microwave-safe bowl. Microwave, uncovered, on HIGH 3 minutes. Stir thoroughly. Microwave, uncovered, on HIGH 3 to 4 minutes, or until thickened. Stir in bananas and maple syrup.

Makes 4 servings.

Education

Choose whole cornmeal to take advantage of the vitamins and minerals supplied by the corn "germ." Degermed or "bolted" cornmeal has had much of the germ-bearing hull removed, along with its valuable store of nutrients. This process does nothing but extend the shelf life of the nutrient-depleted product.

Dried-Fruit Granola

Most granolas are baked with sweeteners and oils. This granola relies on the natural sweetness of dried fruits to add flavor to the grains.

2 cups quick-cooking rolled oats
1 cup plain wheat germ
1/2 cup chopped walnuts
1/2 cup chopped dried figs
1/4 cup chopped dried apricots
1/4 cup chopped pitted dates

1/4 cup golden raisins
1/4 cup sunflower kernels
1 teaspoon ground cinnamon
Fruit juice or milk (optional)
Diced fresh fruit (optional)

Combine oats, wheat germ, walnuts, figs, apricots, dates, raisins, sunflower kernels and cinnamon in a self-sealing plastic bag and store in the refrigerator. Serve with fruit juice, milk or tossed with fresh fruit.

Makes about 4 cups, 8 servings.

Variations

Use any combination of dried fruits to equal 1 to 1 1/2 cups. Use as a hot cereal, placing 1/2 cup granola in a bowl with 3/4 cup water. Microwave, uncovered, on HIGH 2 minutes. Serve with skim milk and brown sugar—delicious!

Education

Wheat germ can provide a nutritional boost when added to breakfasts or baked goods. In addition to being rich in vitamin E, iron, riboflavin and thiamin, the "germ," or growing part of a wheat berry, also supplies protein and fiber. Although it does contain polyunsaturated fat, only 25 percent of its calories are derived from fat, keeping wheat germ well within nutritional guidelines suggesting 30 percent or fewer calories from fat in our diets overall.

Golden Cinnamon Applesauce Oatmeal

Having applesauce on hand is a quick way to add a delicious flavor boost to breakfast.

3/4 cup apple juice, cider or water
1/2 cup quick-cooking rolled oats
1/4 cup applesauce
1 tablespoon golden raisins

1 teaspoon honey
Dash of cinnamon
Dash of salt

Combine juice, cider or water, oats, applesauce, raisins, honey, cinnamon and salt in a microwave-safe bowl. Microwave on HIGH 2 minutes, or until creamy. Stir and serve.

Makes 1 serving.

Rice Porridge with Apricots

One of the wonderful things about quick-cooking grains is that they're perfect for breakfast. Apricots liven up rice for the morning meal.

1/3 cup quick-cooking brown rice
1/2 cup apple juice or water
1 tablespoon chopped dried apricots

1 teaspoon brown sugar
Skim milk (optional)

Combine the rice, juice or water, apricots and sugar in a small saucepan and bring to a boil. Reduce heat, cover and simmer 5 to 10 minutes. Stir, cover and let stand 5 minutes. Serve with milk, if desired.

Makes 1 serving.

Note Cooking times for quick brown rice can vary among brands. Check package directions.

Variation

Substitute raisins, golden raisins or chopped, dried prunes for apricots.

Nutty Banana Cream of Wheat

A favorite combination on sandwiches, peanut butter and banana finds a new place in hot cereal.

1 1/4 cups water
3 tablespoons Cream of Wheat
    cereal
1 teaspoon peanut butter

1/2 banana, diced
1 teaspoon honey
Skim milk

Combine water, cereal and peanut butter in a small saucepan and bring to a boil. Reduce heat and cook, stirring, 2 to 3 minutes, or until creamy.

Add banana and honey and stir. Place in a bowl and add milk.

Makes 1 serving.

Education

Serve cereal with fresh fruit for added fiber. The more refined hot cereals have had much of their fiber removed. By adding fruit, you not only compensate, but add significant vitamins and minerals to your breakfast, too.

Desserts & Delights

Desserts can be an important part of vegetarian cuisine. A variety of protein-, vitamin- and mineral-rich ingredients will add valuable nutrition to your daily diet.

Take nuts, for example. In addition to protein, nuts are one of the best vegetable sources of vitamin E. They are also rich in B vitamins and contain magnesium, zinc, copper and selenium. Because they are high in fat, nuts should be used in moderation. Sprinkle chopped nuts on fruit desserts, add to cookie mix or include nuts in cakes and pies.

Oats, delicious in bar cookies and fruit crisps, have more protein than bulgur wheat and about one and one-half times more protein than brown rice. Oats also come with high levels of iron and zinc, both valuable minerals, and vital in a diet that does not include meats.

Fruits, while not high in protein, do provide fiber (especially apples and pears) and can be good sources of vitamin C (pineapple, berries and citrus fruits) and beta carotene (apricots and melons). Bananas and pears are rich in potassium. Berries and dried fruits provide iron.

Desserts, with nutritious ingredients, not only can provide a delicious ending to a meal, but can help maintain a balance of health-enhancing nutrients and fiber.

Mixed Grilled Fruits with Berry Dipping Sauce

Toss some fruit on the grill for dessert. Keep a careful eye, since it won't take long until dessert is ready!

4 bananas
4 (1/2-inch-thick) slices fresh
    pineapple
2 kiwi fruit, peeled and thickly sliced
8 large strawberries, stems removed
Powdered sugar, for garnish
Mint leaves, for garnish

BERRY DIPPING SAUCE
1 cup whole strawberries, caps
    removed
1/4 cup red raspberries
2 tablespoons maple syrup
1/4 teaspoon vanilla extract
1/8 teaspoon salt

Preheat grill. Slit banana peels along inside curve to prevent bursting. Arrange pineapple slices on rack over hot coals. Place bananas toward outside. Turn pineapple slices and bananas after 3 to 5 minutes and place kiwi slices and strawberries on grill. Cook 2 to 3 minutes or until all fruit is lightly toasted.

To serve, peel bananas and place in small bowls, curved sides up, and dust with powdered sugar. Arrange pineapple slices, kiwi fruit and strawberries on a serving plate. Dust with powdered sugar and garnish with mint leaves. Prepare sauce and serve on the side.

Makes 4 servings.

Berry Dipping Sauce

To make sauce, combine strawberries, raspberries, maple syrup, vanilla and salt in a blender and process until smooth. Pour through a sieve and serve in small, individual bowls.

Golden Carrot Cake
with Pineapple Icing

Golden raisins and pineapple give a glow to a delicious, generously frosted carrot cake. The cake and icing are both dairy-free.

1 cup unbleached all-purpose flour
1/2 cup whole-wheat pastry flour
2 teaspoons baking powder
2 teaspoons ground cinnamon
1 teaspoon ground cloves
1/2 teaspoon ground allspice
1/2 teaspoon freshly grated nutmeg
1/2 teaspoon salt
3 eggs
1/2 cup fresh orange juice
1/2 cup crushed pineapple, drained
1 cup packed light brown sugar
1/3 cup canola oil
2 cups shredded carrots
1 cup golden raisins
1 cup walnuts, chopped

PINEAPPLE ICING
1/4 cup soy margarine, softened
1 1/4 cups powdered sugar
1 teaspoon fresh lemon juice
1/2 teaspoon vanilla extract
1/8 teaspoon salt
1/3 cup packed crushed pineapple,
    well drained

Preheat oven to 350F (175C). Lightly oil a 13 x 9-inch nonstick baking pan.

Sift together the flours, baking powder, cinnamon, cloves, allspice, nutmeg and salt into a large bowl.

Place the eggs, orange juice, pineapple, brown sugar and oil in a food processor and process until smooth, scraping down the sides of the container as necessary.

Stir carrots, raisins, walnuts and blended pineapple mixture into the dry ingredients just until combined. Pour batter into prepared pan and smooth evenly with a spatula. Bake 35 to 40 minutes, or until a wooden pick inserted in the center comes out clean. Let cool in pan. Turn out on a platter before frosting, if desired.

Prepare icing and spread on cooled cake. Cut cake into squares.

Makes 12 servings.

Pineapple Icing

Place the margarine in a medium bowl and beat with a wooden spoon until creamy. Gradually sift powdered sugar into bowl, beating until creamy after each addition.

Beat in lemon juice, vanilla and salt. Add pineapple and beat until combined. Let stand 5 minutes. Beat again and spread on cooled cake.

Creamy Raisin Rice Pudding

Slow baking gives unparalleled flavor to authentic rice pudding. If you've never enjoyed the real thing, dig in. Low-fat milk keeps fat content within reason, but the taste is out of this world. The recipe is simplicity itself, and if you don't like the raisins, just leave them out.

4 cups low-fat milk	1/2 cup raisins
3 tablespoons white rice (not instant or converted)	5 tablespoons sugar
	1/2 teaspoon salt

Preheat oven to 300F (150C). Lightly butter a deep casserole dish.

Combine milk, rice, raisins, sugar and salt in buttered casserole dish. Bake 3 1/2 hours, uncovered, stirring every 15 minutes for the first hour so the rice doesn't settle.

Makes 6 servings.

Fresh Banana Custard Pie

A meringue topping highlights this nourishing banana custard pie.

1 (9-inch) Pie Crust (see page 237)
2 cups low-fat milk
4 egg yolks
1/2 cup sugar
3 tablespoons unbleached
    all-purpose flour
1/4 teaspoon salt
1 teaspoon vanilla extract
2 bananas, thinly sliced

MERINGUE
5 egg whites
1/3 cup sugar
1/4 teaspoon salt

Prepare crust as directed on page 237. Place the milk in a medium saucepan over medium heat and remove when milk is scalded or small bubbles appear around the edge of the pan.

Beat the yolks, sugar, flour and salt together in a bowl. Very slowly add hot milk, stirring constantly. When all of the milk is added, return mixture to the pan and cook over medium heat, stirring constantly, until thickened, about 10 minutes.

Remove pan from heat and pour custard into a bowl. Stir in vanilla. Cover with a sheet of waxed paper pressed against the surface of the custard to prevent a skin from forming. Chill.

To make meringue, place egg whites in a large bowl over a pan of hot water. Stir in sugar and salt until dissolved. Remove bowl from hot water and beat with an electric mixer until the whites are stiff and shiny.

Preheat broiler. Line the bottom of the baked 9-inch pie crust with sliced bananas. Pour chilled custard over bananas and smooth with a spatula. Cover custard with meringue, spreading it evenly to touch the inner edge of the crust all the way around. (This keeps the meringue from shrinking.) Place the pie in middle of oven and broil until lightly browned, 4 to 5 minutes, watching meringue closely to prevent burning.

Makes 6 servings.

Education

To prevent the growth of harmful bacteria, eggs must be kept chilled. If you are setting aside egg whites, as above, for example, cover and place them in the refrigerator. As they whip up better when they are warm, the hot water bath, as above, will help. Always refrigerate egg dishes, such as this pie, and refrigerate any leftovers promptly.

Honeyed Strawberries

For such a simple dessert, only the season's best strawberries will do. Serving at room temperature allows all of the strawberry flavor to come through. This is a gorgeous dessert!

2 pints fresh strawberries
1 tablespoon honey

Mint sprigs, for garnish

Rinse and dry strawberries, keeping stems and caps in place. Place berries in a bowl and drizzle with honey. Let stand 1 hour at room temperature.

To serve, place berries in individual glass dessert bowls. Garnish with mint sprigs.

Makes 4 servings.

Apple Crisp with Cranberries

Colorful and tart cranberries give a tangy twist to a traditional crisp—a dessert you and your guests will love!

3 large, tart baking apples
1/4 cup cranberries, finely chopped
1/4 cup raisins
1/4 cup plus 1 tablespoon whole-wheat pastry flour
1 tablespoon granulated sugar
1 1/2 teaspoons ground cinnamon
1/4 cup apple cider, apple juice or water

1/2 cup rolled oats
1/4 cup finely chopped walnuts or pecans
1/4 cup packed light brown sugar
3 tablespoons canola oil
1/4 teaspoon salt

Preheat oven to 350F (175C).

Peel, core and dice apples. Place apples in a large bowl and add cranberries, raisins, the 1 tablespoon flour, granulated sugar and 1/2 teaspoon of the cinnamon. Toss until combined. Pour cider, juice or water into bottom of baking dish. Add apple mixture in an even layer.

Mix together oats, remaining 1/4 cup flour, nuts, brown sugar, the remaining 1 teaspoon cinnamon, oil and salt in same large bowl. Gently press topping over apples. Bake 45 to 60 minutes, or until apples are tender and topping is crisp and lightly browned.

Makes 4 or 5 servings.

Daphne's Raisin Rock Cookies

Visiting Daphne in Kew, home of the famous gardens, was always an occasion for properly brewed tea and freshly baked rock cookies. You'll love this variation, which is dairy and egg free.

1/2 cup raisins
1 cup water
1/2 cup soy margarine, softened
2/3 cup packed light brown sugar
1 cup old-fashioned rolled oats
1/3 cup chopped walnuts
3/4 cup unbleached all-purpose
　　flour

1/4 cup whole-wheat pastry flour
3/4 teaspoon ground cinnamon
1/2 teaspoon baking powder
1/2 teaspoon salt
1/4 teaspoon freshly grated nutmeg

Preheat oven to 350F (175C). Lightly oil 2 nonstick baking sheets.

Combine raisins and water in a small saucepan and bring to a boil over medium heat. Reduce heat and simmer, uncovered, 5 minutes. Set aside to cool.

Beat together the margarine and sugar in a large bowl. Stir in rolled oats and walnuts. Drain raisins, reserving 1/3 cup of the raisin water. Add the reserved raisin water and raisins to the oat mixture.

In a medium bowl, combine the flours, cinnamon, baking powder, salt and nutmeg. Add to the oat mixture, stirring just until combined. Drop by tablespoonfuls onto prepared baking sheets. Bake 15 to 18 minutes, or until lightly browned. Cool on wire racks.

Makes about 36 cookies.

Granola Squares

A food processor makes it easy to prepare these delicious dessert or snack bars.

3/4 cup raisins
1/4 cup golden raisins
1/4 cup dried cherries
1/4 cup packed light brown sugar
1 cup rolled oats

3/4 cup walnuts or pecans
1/4 cup whole-wheat pastry flour
1/2 teaspoon baking powder
1/4 cup fresh orange juice
1 egg

Preheat oven to 350F (175C). Lightly oil a nonstick 8-inch-square baking pan.

Place all ingredients in a food processor and process until ground to a paste. Press into the bottom of prepared baking pan. Bake 35 to 40 minutes, or until lightly browned. Cut into bars while warm.

Makes 16 servings.

Variations

Chop fruits and nuts finely with a sharp knife instead of using a food processor, then combine with remaining ingredients in a bowl.

Substitute dried cranberries, dried blueberries, dried apricots or pitted dates for cherries, or for golden raisins.

Hot Spiced Apple Slices

Use the microwave to make a quick and tasty cool-weather dessert.

2 large baking apples, peeled, cored
    and sliced
1 tablespoon fresh lemon juice
1 tablespoon apple juice
1 tablespoon light brown sugar

1/4 teaspoon ground cinnamon
1/4 teaspoon ground cardamom
Dash of ground cloves
Dash of salt

Place apples in a 9-inch-square microwave-safe casserole dish and toss with lemon juice and apple juice.

Combine brown sugar, spices and salt in a small bowl. Sprinkle over apples and toss until combined. Cover dish and microwave on HIGH 6 minutes, stirring twice, or until apples are tender.

Makes 4 servings.

Minted Pineapple Fruit Salad

A dramatic dessert with beautiful colors and shapes, suitable for a buffet table.

1 large pineapple, halved lengthwise
2 (11-ounce) cans Mandarin
    oranges, drained
1 cup fresh blueberries
1 cup sliced strawberries

1 cup diced banana
1 tablespoon fresh lemon juice
2 tablespoons packed brown sugar
1 teaspoon minced fresh mint

Remove pineapple flesh with a small knife, leaving a 1/2-inch-thick shell. Core pineapple and cut flesh into cubes. Place in a large bowl with the orange sections, blueberries and strawberries.

Toss the banana with the lemon juice and brown sugar. Add to the fruit mixture with the mint and toss until well combined. Fill pineapple halves with fruit mixture and arrange on a platter.

Makes 8 servings.

Oven-Roasted Red-Skinned Pears

Do not peel pears, or they will lose their shape during baking. The recipe is so very simple; the results, superb! Substitute regular Bartlett or Bosc pears, if you like.

8 red-skinned pears
Mint sprigs, for garnish

Pomegranate seeds, for garnish

Preheat oven to 350F (175C). Lightly oil a 13 x 9-inch baking dish.

Rinse the pears and place stem end up in prepared baking dish. Bake 50 to 60 minutes, or until the pears are quite tender. Serve on individual plates garnished with mint sprigs and pomegranate seeds.

Makes 8 servings.

Autumn Fruit Puree

The difference between this and a jar of commercial applesauce will be evident from the first bite!

2 firm pears, peeled, cored and
    cubed
2 tart apples, peeled, cored and
    cubed
1/4 cup cranberries, finely chopped

1/4 cup golden raisins
1/4 cup apple cider
3 tablespoons maple syrup
1/4 teaspoon ground cinnamon

Combine the pears, apples, cranberries, raisins, cider, maple syrup and cinnamon in a medium saucepan and bring to a boil over medium heat. Reduce heat, cover and simmer 10 minutes, stirring once or twice.

Simmer, covered, until the fruits are soft, about 3 to 4 minutes. Stir well to break up fruits.

Makes 4 servings.

Pie Crust

An easy, flaky pie crust awaits fresh fruit fillings. The crust can also be made using all unbleached flour.

3/4 cup unbleached all-purpose flour	Dash of cinnamon
1/4 cup whole-wheat pastry flour	1/4 cup canola oil
1/2 teaspoon salt	1 1/2 to 2 tablespoons cold water

Preheat oven to 425F (220C).

Combine the flours, salt and cinnamon in a medium bowl. Lightly stir in oil with a fork until the mixture resembles coarse crumbs. Don't overwork the dough, or it will become tough.

Sprinkle 1 tablespoon of water over the flour mixture and stir lightly, adding remaining water a teaspoon at a time until dough is moist and clings together. Shape dough into a ball.

To roll, wipe kitchen counter with a damp cloth. Place a square of waxed paper on the dampened surface. Flatten the ball of dough slightly, and center on the waxed paper. Cover dough with another square of waxed paper. Roll pastry out with a rolling pin until it is about 1/8 inch thick and an inch or so bigger all around than a 9-inch pie plate.

Peel off top piece of waxed paper. Invert pie crust over pie pan and gently peel off remaining sheet of waxed paper. Ease crust into sides of pie pan. Trim off any uneven edges, leaving about 3/4 inch of pastry around edges. Fold this in and pinch to form a high, decorative edge.

Prick pie crust all over sides and bottom with a fork to prevent puffing during baking. Bake 15 minutes, or until golden. Cool before filling.

Makes 1 (9-inch) pie crust.

Variations

Roll out into a circle and place over fresh fruit fillings before baking for a cobbler.

Roll and cut out shapes with cookie cutters. Prick with a fork, sprinkle with sugar, and bake, as above, until golden. Use to top cooked fruit dishes, such as Hot Spiced Apple Slices (see page 235) just before serving.

Blueberry Compote

If you can stir, you can make this low, low-fat, fabulously fresh-tasting dessert. In winter, frozen berries can be used, though cooking will take slightly longer.

4 cups blueberries	2 tablespoons cornstarch
1/3 cup water	1 tablespoon fresh lemon juice
1/4 cup maple syrup	4 mint leaves, for garnish

Place the blueberries in a medium saucepan. In a 2-cup measure, combine the water, maple syrup, cornstarch and lemon juice and stir until well combined.

Add liquid to berries and bring to a boil over medium heat, stirring frequently. Reduce heat and cook, stirring constantly, until mixture is thickened, about 3 to 5 minutes. Let cool slightly, then spoon into glass bowls or parfait glasses. Garnish with a mint leaf. Chill, if desired.

Makes 4 servings.

Variation

Use blueberry mixture to fill a baked pie crust (page 237) or purchased graham cracker crust.

Education

Fresh fruit or the blueberry filling above, served in the prebaked pie shell, provides an enticing dessert without eggs and without dairy products. If you have enjoyed a dinner with either eggs or cheese, balance might be provided by this, or another dairy-free, egg-free selection from these pages.

Chocolate-Glazed Frozen Banana Pops

Wonderful for children (and grown-ups, too) when warm weather brings a craving for a frozen dessert. Choose bananas that are not overly ripe.

4 ripe bananas	1/4 cup sugar
1 square unsweetened chocolate, coarsely chopped	1/4 teaspoon vanilla extract
3 tablespoons hot decaffeinated hazelnut coffee	Dash of salt
	Finely chopped almonds (optional)

Peel bananas and cut in half crosswise. Push flat wooden sticks into the flat end of each banana half and place on a lightly oiled baking sheet. Place in a freezer 1 hour or more.

To glaze bananas, place chocolate in a small metal bowl with the hot coffee. Stir until chocolate is melted. Add sugar, vanilla and salt and stir together until smooth. Allow to cool slightly.

Swirl frozen banana halves in chocolate, tilting the bowl to make covering the bananas easier. Stir chopped nuts into chocolate before coating bananas, if desired. Serve at once.

Makes 8 servings.

Variation

Use chocolate glaze for grilled fruits or for fresh strawberries, fresh banana chunks or orange sections.

Sweet Potato Pie

Aromatic spices and a touch of molasses are combined with the natural sweetness of sweet potatoes to make a delicious pie for the holidays, or any time.

2 large (about 1 3/4 lbs.) sweet
   potatoes
2 eggs, beaten
1 1/2 cups low-fat milk
3/4 cup packed light brown sugar
2 tablespoons medium molasses
1/2 teaspoon salt
1 teaspoon ground cinnamon
1/4 teaspoon ground allspice
1/8 teaspoon ground cardamom

CRUST
1 cup unbleached all-purpose flour
1/2 cup whole-wheat pastry flour
1/4 teaspoon salt
1/8 teaspoon freshly grated nutmeg
6 tablespoons soy margarine or
   butter
3 to 4 tablespoons cold water

Trim ends from sweet potatoes and pierce several times with a sharp knife. Put sweet potatoes into a casserole dish and cover with a lid. Microwave on HIGH 15 minutes, turning sweet potatoes once, or until they are tender. Let cool slightly.

Peel sweet potatoes and mash pulp in a large bowl. There should be about 3 cups. Add eggs, milk, sugar, molasses, salt, cinnamon, allspice and cardamom. Beat until well blended or process, half at a time, in a food processor until smooth.

Preheat oven to 425F (220C). Prepare crust.

Spoon filling into pie crust and smooth top. Bake 10 minutes, then turn heat down to 325F (165C) and bake about 1 to 1 1/4 hours, or until the filling has puffed up and is cooked throughout. Let cool before slicing.

Makes 8 servings.

Crust

Combine flours, salt and nutmeg in a medium bowl. Use a pastry blender to combine margarine or butter and flour until the mixture resembles coarse meal. Stir in enough water so that mixture just holds together. Gather dough together to form a ball. Slightly flatten, then roll out ball on a lightly floured work surface until the dough is 11 to 12 inches in diameter. Fit dough into a deep 10-inch pie plate. Trim off excess dough, leaving about 3/4 inch of pastry around edges, and flute with your fingers or press with the tines of a fork.

Makes 1 (10-inch) crust.

Fruitcake Ring with Macadamia Nuts

Forget the heavy fruitcakes synonymous with holiday giving. This moist and flavorful cake is worlds apart . . .

1 cup packed light brown sugar
4 tablespoons soy margarine or
    butter, softened
1 cup raisins
1 large apple, peeled, cored and
    finely diced
1 cup fresh orange juice
1 teaspoon ground cinnamon
1/2 teaspoon freshly grated nutmeg

1/4 teaspoon ground cloves
1 1/2 cups unbleached all-purpose
    flour
1 tablespoon cornstarch
1 teaspoon baking soda
1/2 teaspoon salt
1/3 cup chopped macadamia nuts,
    walnuts or pecans
Powdered sugar

Preheat oven to 350F (175C). Lightly oil a nonstick tube or Bundt pan with nonstick cooking spray.

Beat together the brown sugar and margarine in a medium bowl until combined. Stir in raisins, apple, orange juice and spices.

Combine the flour, cornstarch, baking soda, salt and nuts in a large bowl. Add raisin mixture to the dry ingredients and stir until combined. Turn batter into prepared pan.

Bake about 50 minutes, or until a wooden pick inserted into the cake comes out clean. Cool in pan 10 minutes, then carefully turn out onto a rack. Dust with powdered sugar when cake is cool.

Makes 12 servings.

Jam Dots

For a colorful assortment of cookies, use a variety of jams if you prefer.

1 cup unbleached all-purpose flour
1 cup whole-wheat pastry flour
1 cup blanched almonds, ground
   (see Note, below)
1/4 teaspoon salt
1 1/4 cups sugar
10 tablespoons butter

2 eggs, beaten
1 teaspoon vanilla extract
1 teaspoon grated lemon peel
1/4 cup strawberry or raspberry
   jam
Powdered sugar, for garnish

Preheat oven to 375F (190C). Lightly oil 2 nonstick baking sheets.

Combine flours, ground almonds and salt in a large bowl. Beat together sugar and butter in a medium bowl until combined. Add eggs, vanilla and lemon peel. Add egg mixture to dry ingredients and stir until combined.

Using your hands, roll dough into balls the size of small walnuts. Place on prepared baking sheets about 2 inches apart. Gently press your thumb in the center of each ball to make an indentation, holding your fingers around the dough to keep the dough from splitting.

Place about 1/4 teaspoon jam in the center of each cookie. Bake 15 to 18 minutes, or until golden. Remove to a rack to cool. Dust with powdered sugar.

Makes about 48 cookies.

Note To blanch almonds, pour boiling water over the shelled nuts. Let stand about 30 seconds. Run under cold water. Drain. Slip off the brown skins with your fingers. Allow nuts to dry before grinding in a food processor.

Walnut Spice Cake

A delicious combination of flavors, delicately spiced and finished off with a crisp walnut topping.

2 cups unbleached all-purpose flour
1 tablespoon unsweetened cocoa
    powder
2 teaspoons baking powder
2 teaspoons ground cinnamon
1/2 teaspoon ground cloves
1/2 teaspoon ground allspice
1/4 teaspoon ground nutmeg
1/8 teaspoon ground cardamom

1/2 teaspoon salt
1 cup raisins
1 1/4 cups granulated sugar
6 tablespoons butter or soy
    margarine, softened
2 eggs, beaten
1/2 cup applesauce
3/4 cup walnuts
2 tablespoons packed brown sugar

Preheat oven to 350F (175C). Lightly oil a 13 x 9-inch nonstick baking pan.

Sift flour, cocoa, baking powder, cinnamon, cloves, allspice, nutmeg, cardamom and salt into a large bowl. Add raisins and stir to combine.

Beat together granulated sugar and butter in a medium bowl until combined. Add eggs and applesauce and beat until smooth. Add to dry ingredients and beat until combined.

Pour batter into prepared baking pan and smooth to an even layer. Combine nuts and brown sugar in a food processor and finely chop. Sprinkle mixture evenly over cake. Bake 40 to 45 minutes, or until a wooden pick inserted in the center comes out clean. Let cool in pan on a wire rack.

Makes 16 servings.

Metric Conversion Charts

Comparison to Metric Measure

When You Know	Symbol	Multiply By	To Find	Symbol
teaspoons	tsp	5.0	milliliters	ml
tablespoons	tbsp	15.0	milliliters	ml
fluid ounces	fl. oz.	30.0	milliliters	ml
cups	c	0.24	liters	l
pints	pt.	0.47	liters	l
quarts	qt.	0.95	liters	l
ounces	oz.	28.0	grams	g
pounds	lb.	0.45	kilograms	kg
Fahrenheit	F	5/9 (after subtracting 32)	Celsius	C

Fahrenheit to Celsius

F	C
200–205	95
220–225	105
245–250	120
275	135
300–305	150
325–330	165
345–350	175
370–375	190
400–405	205
425–430	220
445–450	230
470–475	245
500	260

Liquid Measure to Milliliters

1/4 teaspoon	=	1.25 milliliters
1/2 teaspoon	=	2.5 milliliters
3/4 teaspoon	=	3.75 milliliters
1 teaspoon	=	5.0 milliliters
1 1/4 teaspoons	=	6.25 milliliters
1 1/2 teaspoons	=	7.5 milliliters
1 3/4 teaspoons	=	8.75 milliliters
2 teaspoons	=	10.0 milliliters
1 tablespoon	=	15.0 milliliters
2 tablespoons	=	30.0 milliliters

Liquid Measure to Liters

1/4 cup	=	0.06 liters
1/2 cup	=	0.12 liters
3/4 cup	=	0.18 liters
1 cup	=	0.24 liters
1 1/4 cups	=	0.3 liters
1 1/2 cups	=	0.36 liters
2 cups	=	0.48 liters
2 1/2 cups	=	0.6 liters
3 cups	=	0.72 liters
3 1/2 cups	=	0.84 liters
4 cups	=	0.96 liters
4 1/2 cups	=	1.08 liters
5 cups	=	1.2 liters
5 1/2 cups	=	1.32 liters

Index

THE
PRACTICAL
ENTOMOLOGIST

Above A shield-bug (*Tectocoris diophthalmus*)

THE
PRACTICAL
ENTOMOLOGIST

Rick Imes

A FIRESIDE BOOK
Published by Simon & Schuster Inc.
New York • London • Toronto • Sydney • Tokyo • Singapore

A QUARTO BOOK

F

Simon and Schuster/Fireside
Simon & Schuster Building
Rockefeller Center
1230 Avenue of the Americas
New York, N.Y. 10020

Designed and produced by Quarto Publishing plc, The Old Brewery, 6 Blundell Street, London N7 9BH

Senior Editor Sally MacEachern
Editor Diana Brinton
Designer Neville Graham
Illustrator Wayne Ford
Index Connie Tyler
Picture Manager Sarah Risley
Assistant Art Director Chloë Alexander

Art Director Moira Clinch
Publishing Director Janet Slingsby

Typeset by Bookworm Typesetting, Manchester
Manufactured by J Film Process
Singapore Pte Ltd
Printed in Hong Kong by Leefung-Asco Printers Ltd, Hong Kong

1 2 3 4 5 6 7 8 9 10 Pbk

Library of Congress Cataloging-in-Publication Data
Imes, Rick.
    The practical entomologist / Rick Imes.
        p.   cm.
    Includes index.
    ISBN 0-671-74696-0 (cloth).–ISBN 0-671-74695-2
    (Fireside : pbk.)
    1. Entomology.   2. Insects.   3. Insects –
    Identification.
    I. Title.
    QL463.I44   1992
    595.7–dc20
                                                91-23905
                                                    CIP

DEDICATION

The Practical Entomologist is dedicated to the memory of Robert Wilson Michler of Easton, Pennsylvania, 1919–91.

CONTENTS

THE BASICS OF ENTOMOLOGY

ENTOMOLOGY IN ACTION

DRAGONFLIES & DAMSELFLIES
ORDER ODONATA

GRASSHOPPERS, KATYDIDS, & CRICKETS
ORDER ORTHOPTERA

FOREWORD

Entomology, that branch of zoology that deals with the study of insects, is a science often neglected by amateur naturalists – those whose interest in nature constitutes more of a pastime than a vocation. Most of us relate better to birds and mammals because these warm, furry or fluffy creatures embody traits we deem pleasant, such as parental care, live birth (among mammals), and the melodious songs of many birds. Even wild flowers, with their pleasing colors and agreeable fragrances, fare better than most insects when competing for our favor.

Insects, in contrast, with their six legs, hardened body, bulging compound eyes, and antennae, may at first seem more like miniature aliens from a science fiction movie than fellow earthlings. Some insects sting, some destroy our crops, and a few are parasitic or transmit diseases to people or other animals. Many others, however, are extremely beneficial to us, and insects as a group are vital components of the world's food webs and nutrient cycling. Despite our differences with insects, evolutionary evidence strongly suggests that all life on earth had a common origin and that insects merely constitute a different branch on our own evolutionary tree.

Dominant life forms

In fact, a reasonably strong argument can be made for saying that insects, not mammals, are the dominant life form on earth. Consider that the number of insect species is greater than the number of all other species of organisms combined; beetle species alone outnumber all plant species in the world. Despite their conspicuous scarcity in marine environments, insects inhabit every other conceivable habitat, sometimes in mind-boggling densities. Then consider the remarkable resiliency of insects – their short life-spans and tremendous reproductive capacity result in their being extraordinarily responsive to environmental changes that would stress the populations of other organisms.

The Practical Entomologist is designed to introduce you to a fascinating field of natural history and to dispel any aversions to insects that you may have developed. You will find out about many of the insect orders and families and the differences that set each one apart from the others. Together, we will probe the world of insects, examining some of the unique aspects of their lives. The hands-on activities presented throughout the text provide opportunities for you to get personally involved in entomology and make some of your own observations, rather than just reading about someone else's.

Left The world of insects is rich in brightly colored species. Some colors evolved for defense, others for mimicry or for sexual display. This is an African dragonfly (*Trithemis tirbyi*) in typical posture, waiting predaceously on its perch.

Right The wool-carder bee (*Anthidium manicatum*) cards or braids pieces of moss with its legs and mandibles to form a nest. Here, it is collecting pollen for the nest.

THE BASICS OF ENTOMOLOGY

WHAT ARE INSECTS?

Just what are insects, anyway? Often, any small creature with more than four legs is indiscriminately labeled a "bug," but true bugs represent only one of many different groups of insects. What's more, many of these creepy, crawling critters are not insects at all, but may belong to one of several related but very different groups.

Insects, as it turns out, are characterized by several easily recognized traits that set them apart from any other group of organisms. Like other members of the Phylum Arthropoda (which, literally translated, means "jointed foot"), and unlike mammals, for example, insects possess an external skeleton, or *exoskeleton*, which encases their internal organs, supporting them as our skeleton supports us and protecting them as would a suit of armor on a medieval knight. Unlike other arthropods, their body is divided into three distinct regions – the *head*, *thorax*, and *abdomen*. Insects are the only animals that have three pairs of jointed legs, no more or less, and these six legs are attached to the thorax, the middle region of the body.

Most insects possess two pairs of wings, which are also attached to the thorax; the major exception to this rule are the flies, whose second pair of wings is reduced to tiny vestigial appendages that function as stabilizers in flight. Wings, when present, are a sure indicator that an arthropod belongs to the insect class. However, most ants and a number of more primitive insect groups are normally wingless, so the absence of wings does not by itself mean that the creature in question is not an insect.

Adaptability

It was proposed in the foreword that insects could be considered the dominant form of life on earth. Insects have discovered the basic premise that there is strength in numbers. Their life cycles are quite short, less than one year in most cases, and many have a much shorter span, either by design or through predation. They compensate for this by producing astronomical numbers of offspring; so many, in fact, that were it not for the world's insect-eating animals we would surely be overrun within a very short time.

Short lifespans and high reproductivity arm insects with their greatest advantage – adaptability. It works like this: mutations, those genetic variations resulting in physical, biological, or behavioral changes, occur randomly in every population of organisms. When large numbers of offspring are produced, mutations are therefore relatively frequent, and some invariably enhance an individual's ability to compete for its needs or to adjust to changes in its surroundings. Beneficial mutations afford better odds of reaching sexual maturity and passing on the advantageous trait to future generations. Thus equipped, such "improved" individuals can rapidly replace large segments of their species' population that have been decimated by some disturbance in their surroundings.

Top right Some insects, such as this stick insect from San Cristobal in the Soloman Islands, have elongated bodies. Its twiglike form and nocturnal habits allow the insect to move through vegetation virtually undetected. The surface of its integument is warty and armed with a row of spines.

Ants, being social insects, are very industrious. (**Bottom left**), black ants (*Lasius niger*) in a formicarium caring for pupae and larvae. (**Bottom right**), the colorful red skimmer (*Libellula saturata*), from Mexico, rests on a twig. Prominently displayed are the three body parts: head, abdomen and thorax.

CLOSE RELATIVES

There are several groups of animals that could possibly be confused with insects, and all of these are members of the group known as arthropods. Arthropods compose most of the known animal species, and about 800,000 of the 900,000 or so species of arthropods are insects. The others include crustaceans, spiders, centipedes, and millipedes.

Exoskeletons

All have exoskeletons containing varying amounts of *chitin*, a durable organic compound. It was once thought that the amount of chitin present determined the rigidity of the exoskeleton, but more recent research showed that its hardness is proportional to the protein content of the outer layer, or *cuticle*, and that more chitin is found in the soft inner cuticle. In addition to providing protection against injury, the exoskeleton is very water resistant, which inhibits water loss through evaporation. This major evolutionary adaptation allowed arthropods to colonize dry land while other invertebrates were restricted to aquatic habitats.

For all of its advantages, the exoskeleton of an arthropod is also a hindrance. Its weight limits the maximum size that any arthropod may attain, so none becomes very big and the largest are invariably aquatic, where buoyancy helps offset the greater burden. The non-elastic nature of the exoskeleton's outer cuticle is an obstacle to growth, for in order to attain a larger size, hard-shelled arthropods must first shed, or *molt*, their outer layer, which splits open along a genetically-determined seam. Through this opening emerges the now soft-bodied animal, whose elastic inner cuticle can accommodate growth. Those arthropods that rely upon a very hard exoskeleton for defense are particularly vulnerable at this time and often hide until their growth period is over and their armor has again hardened. Most arthropods molt from four to seven times throughout their life.

Also common to all arthropods are bodies that are segmented to varying degrees, jointed appendages (some of which have differentiated to perform specialized functions), and relatively large and well-developed sensory organs and nervous systems, which enable the animals to respond rapidly to stimuli.

Crustaceans

Named for the Latin term *crusta*, meaning "hard shell," nearly all crustaceans are aquatic, and most live in marine environments, although a few of the most familiar, such as crayfish and water fleas, inhabit freshwater, while others, such as certain species of crab, are to be found in brackish water. Lobsters, fairy shrimp, and barnacles are well-known marine crustaceans; sowbugs, those small armored creatures one finds under rocks or in soil, are among the few terrestrial crustaceans.

The head and thorax of crustaceans are combined into one structure, the *cephalothorax*, which may be covered by a shieldlike *carapace*. Their number of paired appendages is variable, but they have at most only one pair per body segment. Only some of these are "legs," attached to the cephalothorax and used for walking. In some species, the first pair of legs are equipped with large pincers modified for grasping offensively or defensively. Other appendages are variously

Top Like insects, crayfish are also arthropods, but they belong to an entirely different class.

Above The shield-like carapace of this crab covers its cephalothorax and is typical of the hard shell that gives crustaceans their name.

adapted for different functions, such as equilibrium, touch, and taste, chewing, food handling, mating, egg-carrying, swimming, and circulating water over the gills.

Some crustaceans are so unusual that their membership in the Class Crustacea can only be determined in their larval stages by zoologists. The barnacles that tend to encrust any marine surface and the water fleas commonly used in high school biology lab experiments are two such oddballs.

Horseshoe crabs
Though unlikely to be mistaken for any type of insect, these "living fossils" are nonetheless arthropods, and the two groups share some very basic features. Horseshoe crabs, named for the shape of their brown, domed carapace, are marine animals. There are two prominent compound eyes, located atop the carapace, as well as two inconspicuous simple eyes. They have a dorsal abdominal shield edged with short spines, and a bayonetlike tail that, despite its formidable appearance, functions mainly to turn the beast over after it has been flipped upside-down by the surf, lest it remain stranded out of water or succumb to ravenous gulls. Horseshoe crabs have six pairs of jointed appendages on the cephalothorax.

Spiders and their kin
Members of the Class Arachnida (from the Greek term for spider, *arachne*) include spiders, scorpions, ticks, mites, and others. It is this group more than any other that is usually confused with insects. Like crustaceans, the body of an arachnid is divided into a cephalothorax and an abdomen. Arachnids have four pairs of jointed legs, all attached to the cephalothorax, although some, like scorpions, possess a pair of large *pedipalpi*, appendages armed with formidable pincers that may resemble legs but are actually modified mouthparts. They also have one pair of *chelicerae*, mouthparts that, among spiders, each terminate with a fang, at the tip of which is a duct connected to poison glands. Unlike either insects or crustaceans, arachnids have no antennae.

Centipedes and millipedes
The name centipede means "one hundred feet," and centipedes are characterized by having one pair of legs per segment; while few centipedes have exactly one hundred legs, the number is a fair estimate. Their long, flattened, multi-segmented bodies comprise between 15 and 181 segments. The head bears a pair of long antennae, a pair of mandibles for chewing, and two pairs of maxillae for handling food. A pair of poison claws on the first segment behind the head enables a centipede to deliver a painful bite if handled carelessly. Most species live under stones or logs, emerging at night to prey upon earthworms and insects, which they kill with their venomous bite.

The prefix "milli-" means thousand, so does a millipede have one thousand feet? Not really, but one might think so to watch this wormlike creature walk. Each of the 9 to 100 or more abdominal segments sports two pairs of legs, this being the chief difference between millipedes and centipedes. The undulating movement of all these legs as the millipede slowly travels is nothing short of mesmerizing. They avoid light, and live for the most part beneath rocks and rotten logs, scavenging dead plant and animal matter. When threatened, they may roll into a tight ball or a spiral to protect their more vulnerable undersides.

Top Unlike insects, spiders have only two major body segments and four pairs of legs. This perfectly disguised crab spider (*Misumena vatia*) consumes a fly.

Above The many distinct segments and legs of this large millipede (*Sigmoria aberrans*) distinguish it from insects.

TAXONOMY: ORDER FROM CHAOS

Taxonomy is the scientific discipline which puts order into an immensely diverse world and allows scientists to discuss any organism and know with certainty that they are talking about the same species. There are two important divisions of taxonomy. *Classification* is the arrangement of organisms into orderly groups. *Nomenclature* is the process of naming organisms.

Common names are generally used in everyday conversation, but they alone do not positively identify a particular species. Many plants and animals have more than one common name, and are often known by different names in different geographical areas, while the same common name may be assigned to two or more totally different species. Clearly, the potential for confusion is great, with well over 800,000 insect species identified and many more still undiscovered.

Contemporary scientists around the world categorize organisms by means of a classification hierarchy, a system of groupings arranged in order from general to specific relationships. They are, in order of increasing specificity: kingdom, phylum (or division, in the plant kingdom), class, order, family, genus, and species. Each of these is a collective unit composed of one or more groups from the next, and more specific, category. Taking them in reverse order, a genus is a closely-related group of species; a family is an assembly of associated genera; an order is a set of similar families, related orders are combined to form a class, similar classes make up a phylum, and all related phylums constitute a kingdom. The complete classification of a honeybee, for instance, is Kingdom Animalia (animals), Phylum Arthropoda (joint-footed animals), Class Insecta (insects), Order Hymenoptera (bees, ants, and wasps), Family Apidae (bumblebees and honeybees), *Apis mellifera*.

All of the above categories are strictly human concepts, and as such they are subject to differences in interpretation throughout the scientific community, even with such a clear-cut system in place. Among taxonomists, there are the "splitters" and the "lumpers." Splitters are inclined to create many subdivisions among organisms, bas-

Carolus Linnaeus

ing these upon more minute criteria, while lumpers tend to generalize and recognize fewer categories in the same group of organisms. Taxonomy is an active science, and there are occasional changes among accepted classifications that may confuse anyone who does not keep up with scientific literature. In such a case, a glance at the date of the publications containing the questionable terms will indicate which is likely to be the more recent interpretation.

The binomial system

While classification has always been a fairly simple affair, nomenclature has not. By the beginning of the 18th century, the use of Latin in schools and universities was widespread, and it had become customary to use descriptive Latin phrases to name plants and animals. Later, when books began to be printed in different languages, Latin was retained for the technical descriptions and names of organisms. Since all organisms were grouped into genera, the descriptive phrase began with the name of the genus to which the organism belonged. All mints known at that time, for example, belonged to the genus *Mentha*. The complete name for peppermint was *Mentha floribus capitatus, foliis lanceolatis serratis subpetiolatis*, or "Mint with flowers in a head; leaves lance-shaped, saw-toothed, and with very short petioles." The closely related spearmint was named *Mentha floribus spicatis, foliis oblongis serratis*, which meant "Mint with flowers in a spike; leaves oblong and saw-toothed." Though quite specific, this system was much too cumbersome to be used efficiently.

In 1753, Swedish naturalist Carolus Linnaeus introduced a two-word system of naming organisms. This system quickly replaced the older, clumsier method, and came to be known as the Binomial System of Nomenclature (binomial – two names). According to this, individual species are identified by linking the generic name with another word, frequently an adjective. Occasionally, however, the splitters will create two or more subspecies out of what had been a single species, in which case the subspecies name is tacked on after the genus and species, creating a trinomial (three names). All scientific names are Latin, although some have descriptive Greek roots. The first name is always capitalized, but never the second, and both are always either underlined or italicized. When more than one member of the same genus is being discussed, the first name may be abbreviated, as in *D. melanogaster* for *Drosophila melanogaster*.

Phylogenetic tree of life forms

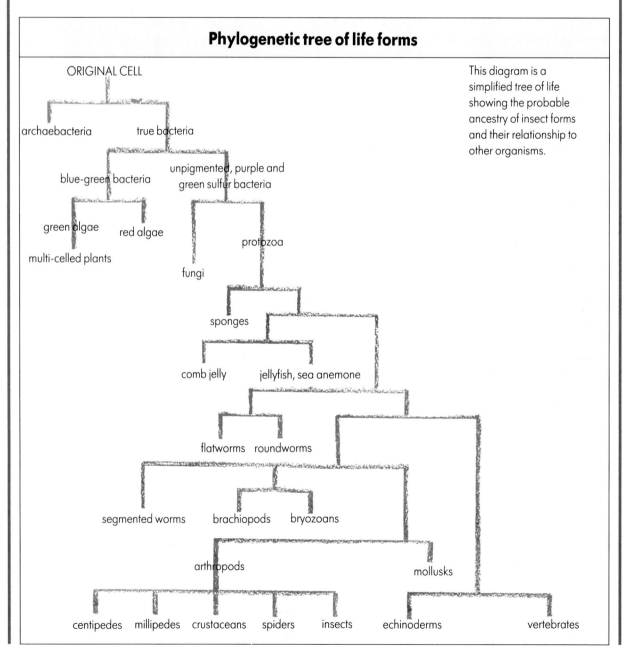

This diagram is a simplified tree of life showing the probable ancestry of insect forms and their relationship to other organisms.

ANATOMY AND MORPHOLOGY

Morphology is the study of external form and structures, the criteria that result in insects being classified as insects and not as something else. Variations on these features define different orders, families, and genera of insects. Related to morphology is anatomy, the internal arrangement of organs and muscles. Learning the basics of both will help you to understand insect lives.

As we mentioned in the beginning of this chapter, the bodies of insects are sheathed in a tough exoskeleton, the hardness of which varies from one species to the next. Because they have no backbone, the support of the exoskeleton is absolutely essential to their mobility on land. The bodies of all insects are divided into three obvious regions – the head, the thorax, and the abdomen.

An insect's head is composed of numerous plates, or *sclerites*, fused together to form a solid capsule that bears one to three simple eyes, two compound eyes, one pair of antennae, and mouthparts. It houses the brain, a fairly simple bundle of nerves from which the nerve cord extends and runs the length of the body along its ventral surface.

The thorax of an insect is divided into three distinct segments. From the head backward, they are the *prothorax*, *mesothorax*, and *metathorax*, each of which is rather box-shaped and composed of four hardened sclerites. The upper (*dorsal*) sclerites of the thorax are called the *notum*, the lower (*ventral*) surface is the *sternum*, and the side (*lateral*) regions are the *pleura* (singular, *pleuron*). Thus, a combination of these terms can isolate any region on the thorax, such as the *pronotum*, *mesosternum*, and so on. A triangular region on the mesonotum, the *scutellum*, is present on all adults, but conspicuous on true bugs (Order Hemiptera).

One pair of legs is attached to each segment of the thorax near the bottom of the pleura. From the thorax outward, the segments of the leg are the *coxa, trochanter, femur, tibia, tarsus,* and *pretarsus*. In addition, adult insects may be wingless or they may have a pair of wings on the mesothorax alone or on both the mesothorax and metathorax.

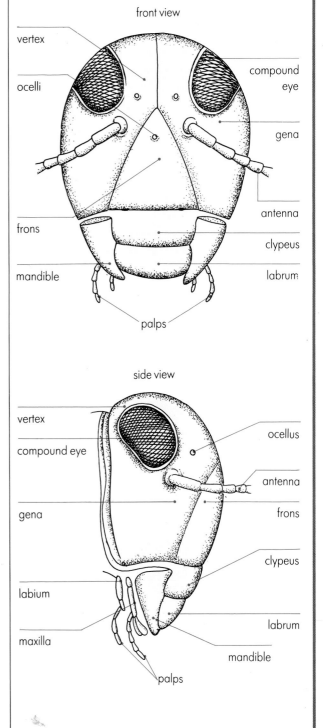

Below Front and side views of an insect's head showing the plates, or sclerites, forming the solid capsule that bears the eyes, antennae and mouthparts.

front view

vertex

ocelli

frons

mandible

compound eye

gena

antenna

clypeus

labrum

palps

side view

vertex

compound eye

gena

labium

maxilla

ocellus

antenna

frons

clypeus

labrum

mandible

palps

Learning the basics of morphology and anatomy will help you to understand more about how insects work and behave. Study a captured insect and compare its structure with the diagram (**below**) of the external form of a grasshopper. Dissect an insect and look for the mouth parts shown in the diagram (**bottom**).

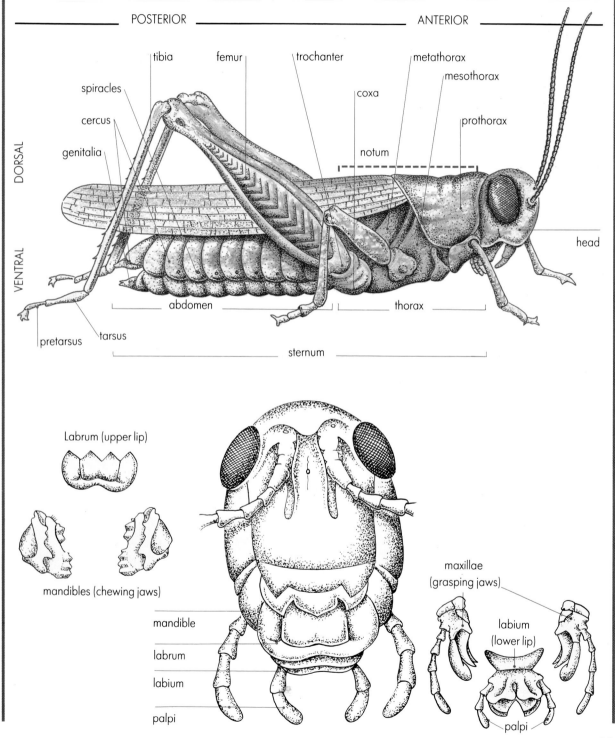

POSTERIOR ——————————— ANTERIOR

tibia femur trochanter metathorax

spiracles mesothorax

coxa

cercus prothorax

DORSAL

genitalia notum

head

VENTRAL

abdomen thorax

pretarsus tarsus

sternum

Labrum (upper lip)

mandibles (chewing jaws)

mandible

labrum

labium

palpi

maxillae (grasping jaws)

labium (lower lip)

palpi

15

The abdomen

The abdomen, which is softer and more flexible than the head or thorax, consists of eleven segments, although some may be reduced in size and not easily visible. The dorsal surface is called the *tergum*; the ventral side is the *sternum*. It is devoid of appendages except for terminal *cerci* of various sizes and shapes and *genitalia*, or reproductive structures. Females may bear an *ovipositor* for egg-laying, and male genitalia may or may not be extended.

The abdomen is necessarily flexible because it houses the *tracheal system* – the breathing apparatus of the insect – and must expand and contract in order to take in and expel air through *spiracles*, which are openings on each side of the abdomen. There is generally one pair of spiracles per abdominal segment, and they lead to a branching network of air tubes, or *tracheae*, and air sacs throughout the body. From these, oxygen can flow to all organs and tissues, and waste gases can be passed out of the body. Normally, anterior spiracles inhale and posterior spiracles exhale.

Circulatory and digestive systems

Unlike that of vertebrates, the circulatory system of insects is completely independent of their respiratory system and is not involved in oxygen transport. The open system is therefore simple, with a tubelike heart that sucks blood in the posterior end and expels it toward the anterior end. The effect is rather like swirling water in a bathtub, and it makes a stark contrast to our own closed circulatory system in which the blood is always enclosed in vessels, no matter how small. Though inefficient by our standards, insect circulatory systems serve their purpose, since the blood transports only food and waste products.

Despite a number of variations, most insect digestive systems are complete, meaning that a closed tube extends from the mouth all the way through the body to the anal opening, where waste is expelled. There are three main regions: the foregut, midgut, and hindgut, variously modified according to the food eaten by that species.

The strength of insects relative to their small size is legendary; ants, for example, are known to be capable of carrying many times their own weight. These remarkable feats are made possible by the special arrangement of muscles, which are attached to the *inside* of their skeletons, affording tremendous leverage. Muscles are basically attached either within individual segments, enabling the insect to expand or contract, or to adjacent segments, allowing the entire body to flex or simply to curl by coordinating the muscles in a series of segments. Joints generally move only in one plane, so that a series of joints oriented in different planes are necessary to give the legs a full range of motion.

Right The tracheal system of a grasshopper. Air is taken in through the anterior spiracles and circulated through a network of trachea and air-sacs, enabling oxygen to flow to all parts of the body. Waste gases are exhaled through the posterior spiracles.

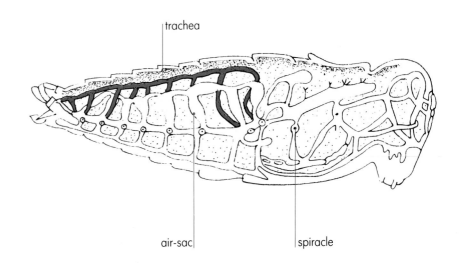

trachea

air-sac

spiracle

What you can do

Dissecting an insect

You will need:
Equipment (see p45)
dissecting tray
2 probes
forceps
X-acto knife
fine scissors
killing jar
hand lens or binocular/
stereomicroscope
pins

Dissecting an insect will allow you to see first-hand the information presented on these pages. The larger the insect, the better – adult grasshoppers are the best and most easily found, but if you can't find a suitable specimen, the biological supply companies listed in the appendix can supply them for a reasonable price.

1 Sacrifice the insect by placing it in a killing jar (see p54) or in the freezer overnight. Place the thawed grasshopper in a dissecting tray. With the hand lens or stereomicroscope, examine the details of its external anatomy and try to locate all of the features shown in the diagrams. Sketching the whole insect or various parts in a field notebook will help you to remember what you see.

2 Carefully stretch out the wings and notice their appearance and texture (you may use pins to hold the wings in position and free your hands if you wish). The tough, leathery forewings protect the more delicate, membranous hind wings. With the X-acto knife, remove the wings after you have finished examining them.

3 Use a probe to open the mouth and examine the mouthparts – are they used for sucking, chewing, or sponging? With forceps and the knife, separate the mouthparts and try to identify the labrum, mandibles, maxillae, and labium. Which parts have palpi? Which are used for grasping, and which for chewing? (see p.15). Slice a thin layer from the surface of one of the compound eyes and examine it with the hand lens or stereomicrosope. What shape are the facets?

4 Place the grasshopper right side up in the dissecting tray. With the knife or scissors, cut a section from the exoskeleton, being careful not to penetrate too deeply, and remove. Find the abdominal organs shown in the illustration below. Flood the

dissecting tray with water; this will help to separate the organs for a better view, and it will also float the tracheal system, which will appear as a mass of silvery, branched tubes.

5 The heart will be on top if it was not cut away with the exoskeleton. A female may have large yellow ovaries obscuring everything else; push these aside with a probe to find the dark-colored stomach and the Malpighian tubules (a waste removal organ equivalent to our kidneys). Follow the stomach toward the mouth to locate the digestive glands and crop, which is a food storage sac. Along the ventral (lower) wall of the abdomen you may find the nerve cord with its enlarged *ganglia*, from which nerves branch to tissues.

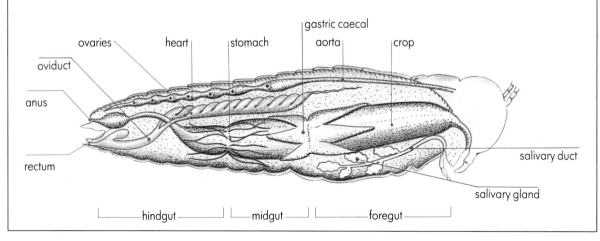

hindgut — midgut — foregut

MOUTHPARTS AND FEEDING

An insect's mouth is composed of distinct parts, each serving a specific function. These mouthparts are a clue to the insect's feeding habits and therefore can tell us much about its life cycle and ecological relationships. Among insects, mouthparts are one means of identification, as they are diversely modified to ingest different types of food, but they all fall into one of several categories.

Chewing mouthparts

Chewing mouthparts are the most common type, and are also the mechanism that most closely resembles that of the human mouth. From front to back, chewing mouthparts consist of the *labrum*, analogous to an upper lip; a rather massive pair of toothed, jawlike *mandibles*, adapted for cutting, crushing, and grinding; a pair of *maxillae*, smaller but also jawlike for grasping, and the *labium*, or lower lip. Each of the maxilla is equipped with an antennalike appendage – the *maxillary palps* – which is used for touching and tasting potential food. The labium bears shorter sensory *labial palps*, and is used to guide food into the mouth cavity. Resting on the labium inside the mouth is a tonguelike *hypopharynx*. Major insect groups with chewing mouthparts include dragonflies, damselflies, grasshoppers, crickets, katydids, and beetles. Many bees combine chewing mouthparts with an elongated labium for lapping fluids, especially nectar.

Sucking mouthparts

The other major type of mouthparts are sucking mouthparts. Whereas insects with chewing mouthparts consume solid food for the most part, sucking insects ingest only liquid food, usually plant juices or body fluids. Sucking mouthparts have been modified into a *proboscis*, or beak, composed of an elongated tubelike labium that sheaths the slender, swordlike mandibles and maxillae, which do the actual piercing; these are called *stylets*, and they enclose the food and salivary channels. When you watch an insect, such as a mosquito, about to bite, you can see the labium bend back in the middle to expose the stylets. After the stylets pierce, saliva is injected

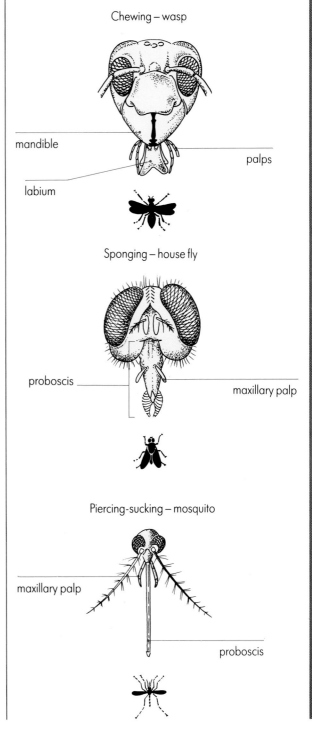

Below The mouthparts of an insect help to identify it, reveal its feeding habits, and indicate what kind of life style it leads. These diagrams show some modifications found within the main categories – chewing, sucking and sponging.

Chewing – wasp

mandible

labium

palps

Sponging – house fly

proboscis

maxillary palp

Piercing-sucking – mosquito

maxillary palp

proboscis

Above Typically, apollos are mountain butterflies, laying their eggs on sedums. Here, an adult (*Parnassius apollo*) feeds on knapweed. The proboscis is extended to enable the butterfly to sip nectar.

through the salivary channel, which causes the subsequent irritation of a mosquito bite, then the food is sucked up through the food channel.

The most common examples of sucking mouthparts are the *piercing-sucking* variety, which are found in the true bugs, leafhoppers, treehoppers, fleas, sucking lice, and some flies. *Lacerating-sucking* mouthparts, found on some flies, are similar, but instead of piercing, the stylets are modified to cut the skin minutely, and the fly sucks the blood that flows from the wound. Butterflies and moths do not pierce or cut; they have *siphoning* mouthparts, their proboscis being coiled like a watch spring under the head when not in use and extended to its full length to sip nectar from flowers.

Sponging mouthparts

Most flies have sponging mouthparts that do not quite fit into any of the above categories. A fleshy labium on the proboscis tip acts like a sponge, extending to soak up liquids and food particles.

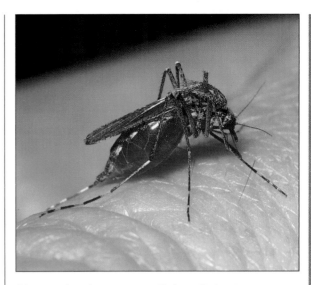

Above A female mosquito (*Culiseta annulata*) fills its abdomen with human blood by means of its syringe-like stylet.

Below Clad in gladiatorial warning colors, especially along its abdomen, this South African grasshopper (*Dictyophorus spumans*) is busy eating a leaf.

What you can do

Predatory games

Equipment:
hand lens
sweep net
examining jar or bug box

If, while collecting, you happen to catch an assassin bug, wheel bug (see p80-7), or any other insect you know to be predatory, you can place it in an examining jar and introduce a likely prey species, such as a fly, in order to witness the piercing-sucking mouthparts in action. A bit of patience is required until both settle down. Of course, the game is rigged in favor of the predator, but you will see nothing that doesn't happen countless times on every acre of land every day during the warmer seasons.

If you're not prepared for such violence, catch a grasshopper and put it in the jar alone. After it settles, put in a crisp piece of lettuce and watch the action of the chewing mouthparts through your hand lens.

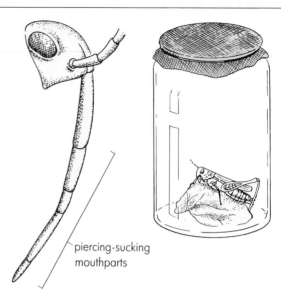

piercing-sucking
mouthparts

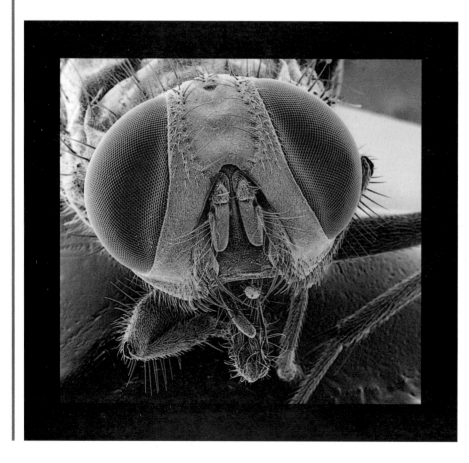

Left The ubiquitous housefly (*Musca domestica*) has been called the most dangerous animal on earth, since it is responsible for transmitting so many diseases to people. One moment a housefly might be on cow dung, the next on sandwiches. Its six feet and sponging mouthparts provide a large surface area with which to transfer pathogens.

WINGS AND FLIGHT

The advantages of flight undoubtedly played a large role in the success of the Class Insecta. Insects were the first creatures on earth capable of flight, which allowed them to more easily escape their enemies, to cover more territory in search of food, water, or mates, and to colonize new areas. They could cross large bodies of water, which were insurmountable barriers to most non-flying terrestrial animals.

Except for flies, all flying insects have two pairs of wings, one of which is attached to the upper mesothorax and the other to the upper metathorax. It is likely that their wings originated as flaps that could be extended from the thorax, allowing wingless insects to escape danger by leaping from an elevated perch and gliding some distance away. Insect wings are unique, having evolved specifically for flight, while the wings of birds and bats are merely modifications of pre-existing limbs.

The earliest insects known to be capable of true flight had two pairs of wings that remained extended and did not fold, even when the creature was at rest. Each pair flapped independently of the other pair, a contemporary parallel to this feature being found in the wings of dragonflies, which are members of a primitive but common order. Many advanced insects, such as the beetles, butterflies, and wasps, have evolved means to link their forewings and hind wings together to form two coordinated flight surfaces rather than four.

Most insect wings are laced with distinct *veins*, the pattern of which is often critical to the identification of individual species. The spaces between the veins are *cells*; those extending to the wing margin are *open cells*, and those enclosed by veins on all sides are *closed cells*. Adult insects that emerge from a pupa have wings that at first look crumpled and useless. Extensions of the tracheal system run through the veins, and blood circulates in the spaces around the tracheae. As air is pumped through the veins, the wings unfurl and straighten. As they harden, veins provide both strength and a degree of flexibility, and the wings become capable of sustaining flight.

Veins tend to be thicker and stronger near the body and along the forward, or "leading" edge, and thinner and more flexible near the tip and along the trailing edge. The trailing edge curls on both

Below It is worth becoming familiar with the terms used to describe the areas and venation of butterfly and moth wings as this will help you to identify a particular species when you are consulting a field guide.

Costal (c)
Subcostal (sc)
Radius (r)
Radial sector (rs)
Medial (m)
Cubitus (cu)
Anal veins (av)
Discal cell (dc)

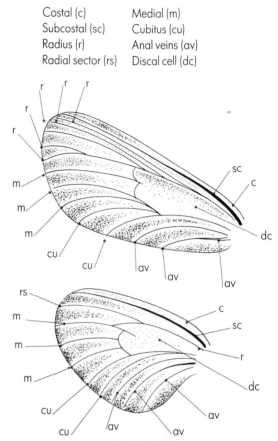

Above Thousands of tiny scales cover the wing membranes of a butterfly, and their varied colors form intricate patterns, as in the yellow bands and eye-spot on the hind wing of this swallowtail (*Heraclides cresphontes*).

Below The wings are lifted as the vertical muscles contract, flattening the thorax (top); then the horizontal muscles contract elongating the thorax (bottom).

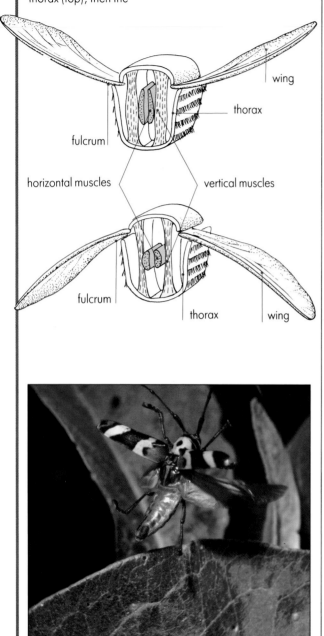

wing

thorax

fulcrum

horizontal muscles

vertical muscles

fulcrum

thorax

wing

Above This longhorn beetle (*Adesmus druryi*), taking off from cloud forest vegetation in Costa Rica, displays its brightly-colored first pair of wings, which are modified as hardened and protective elytra.

the upstroke and the downstroke, pushing against the air behind it and producing not only lift but forward propulsion and reduced drag.

Propulsion

Insect wings do not move simply by muscles pulling at the base, as one might guess. Instead, two different groups of *indirect flight muscles*, housed inside the thorax, work to alternately elongate and flatten the thorax in a vertical direction. The wings, wedged between the upper and lower thoracic sections, move by leverage on a pivotal point, or *fulcrum*. As the vertical muscles contract, the thorax flattens and the wings move up; then the horizontal muscles contract, pulling the sides in, driving the upper and lower thorax higher, and the wings move down. Smaller *direct flight muscles* at the base of each wing adjust the angle of the stroke and therefore the direction of flight.

The frequency of wing beats varies from species to species, from one individual to another, and even in the same individual at different times. Generally speaking, insects such as butterflies, which have large, light bodies and large wings, need far fewer wing beats to stay aloft than do those with small wings and relatively heavy bodies, such as a housefly or a honeybee. Maximum air speed is also highly variable, but is generally less than 20 miles per hour.

Insects, like most other animals, function more efficiently at warmer temperatures. As cold-blooded creatures, insects cannot rely on their body metabolism to generate heat and must use alternative methods to warm themselves. It is not unusual on a chilly day to see such insects as moths and bees rapidly vibrating their wings as they warm up their flight muscles for takeoff. Others will bask in a patch of warm sunlight; most notable among these are many butterflies that spread their wings and orient the surfaces toward the sun's rays, causing them to function as solar collectors.

Wings have evolved to serve a number of other purposes besides flight. Male crickets and katydids, for example, have developed specialized structures on their forewings; when rubbed together on warm summer nights, these structures produce the pulsing songs by which the males seek to attract females. See pages 38–9 for more about wings.

23

LIFE CYCLES

The vast majority of insects lay eggs, and the development of the embryo progresses outside the mother's body. Most species undergo noticeable changes in form as they mature, a process known as *metamorphosis*. Nearly all insects display either *hemimetabolous* (incomplete) metamorphosis or *holometabolous* (complete) metamorphosis, although a few change so little, except in size, that they are said to have *ametabolous* metamorphosis, meaning that there is practically no change in form.

Incomplete metamorphosis

Hemimetabolous insects are usually distinguished by immature stages, called *nymphs*, that resemble adults, the main changes being an increase in size and the development of sexual organs and wings. Nymphs have mouthparts and compound eyes like their adult forms, and eat the same foods. Their wings begin as external pads on the thorax and develop with successive molts. The stage preceding each molt is known as an *instar*, and each succeeding instar more closely resembles the adult stage than did the previous one. Molting allows for growth, as the newer cuticle is more elastic than the old one.

Among certain hemimetabolous insects, specifically dragonflies, damselflies, mayflies, and stoneflies, the immatures, known as *naiads*, are aquatic and do not resemble their terrestrial adult forms.

Complete metamorphosis

The life cycle of holometabolous insects consists of four distinct stages. From the *egg* hatches a *larva*, whose primary functions are to eat and grow. Wormlike in appearance, larvae do not resemble adults; in fact, they scarcely resemble insects. A larva usually possesses a series of simple eyes on its head, though these may be difficult to distinguish. It will also have either chewing or chewing-sucking mouthparts, a pair of very short antennae, and sometimes three pairs of true legs, although there may be other appendages that resemble legs, or they may have no legs at all. Wings, though developing, are hidden under the cuticle. Larvae molt several times to accommodate growth, and each stage preceding a

Below Nearly all insects lay eggs and the embryo then develops outside the mother's body. There are three types of metamorphosis.

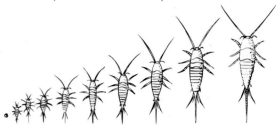

Above A few primitive insects such as silverfish undergo almost no change between the first nymphal stage and the adult form except to increase in size. This is known as ametabolous metamorphosis.

Above Incomplete or hemimetabolous metamorphosis is generally characterized by immature forms called nymphs that resemble adults, or naiads that do not, but gradually increase in size and develop wings and sexual organs.

Left More advanced insects undergo complete or holometabolous metamorphosis. The egg hatches into a larva which goes through a pupal stage before an adult emerges that may be completely different in form, habitat and feeding habits.

Above left The tiny 2mm diameter eggs of the puss moth (*Cerura vinula*) are laid in small groups on the undersides of tree leaves, such as sallow, willow or poplar. They hatch in about a week, depending on temperature.

Above right The mature larva is quite unlike the jet black larvae of the early stage. The zig-zag pattern and the nine pairs of white spiracles are clearly visible. When molested, the larva shows off its red head and waggles a pair of long tails to deter predators.

Bottom left The cocoon of the puss moth is not only well camouflaged but is remarkably tough, protecting the pupa. When the larva is preparing to change into a pupa it chews some bark and incorporates this with its silk to make the cocoon.

Bottom right To escape from the hard cocoon, the moth has a special cutting device on its head. The feathery antennae of the male puss moth are unlike the thinner ones of the female. The puss moth gets its name from the furry appearance of its body.

molt is known as an *instar*.

At the end of the larval stage, a final molt may occur with a *pupa* emerging, or the last larval "skin" may harden into *puparium*. Pupa do not eat and their movement is usually restricted to no more than a wiggle. In this stage, a great transformation is occurring. Some tissues differentiate, others break down and are reabsorbed and reorganized to form new structures. Through the hardened pupal case, the developing wings can often be seen, as can the compound eyes, antennae, mouthparts, and legs. Inside, reproductive organs develop and the digestive system undergoes modifications.

The pupal stage can last from four days to several months, depending upon the species. At *eclosion*, the *adult* emerges, its wings crumpled and its body soft. Within hours, the wings unfurl, becoming stronger as the veins dry and stiffen, and the exoskeleton also dries, hardens, and gains pigment. In most species, adults have a few weeks to accomplish their primary mission of mating and egg-laying, but their tenure in this stage may last either less than two hours, in the case of certain mayfly species, to several years.

Sexual reproduction

Sexual reproduction among insects is the norm; the male of a species transfers sperm to a female, and the sperm are then stored in a special sac in her abdomen. Here the relationship ends, with each of the pair going their separate ways; in some cases, the male dies soon after mating. Egg-laying, or *oviparous*, females are equipped with abdominal appendages called *ovipositors*, which are variously modified to deposit eggs in a site suitable for their development, always close to an appropriate food source. As the eggs are laid, they meet sperm on the way out of the female. Fertilization occurs through a small opening, the *micropyle*, usually shortly after the eggs are deposited.

Among some insects, the eggs remain inside the mother until they hatch. In that case, if the embryo feeds only on material stored inside the egg, it is *ovoviviparous*. In rare instances an embryo can be *viviparous*, being nourished by the mother's tissues prior to hatching.

Once the eggs are laid, they are abandoned by the mother, who usually dies shortly afterward. As compensation for this and for the often vulnerable

What you can do

Observing metamorphosis

It is said that a leopard can't change its spots, but many insects can. The transformation of an insect from larva to adult is a marvelous spectacle, and one that you can observe in your own home. You will need to make a grasshopper zoo (page 77) or other insect cage (page 51), adapting it for whichever species you choose to raise by including the proper food. This means that you must first be reasonably certain of the identity of your specimen and look up its preferred food in a field guide.

Caterpillars of moths and butterflies are relatively easy to identify and feed. With proper care, you will need to collect only one or two individuals to observe metamorphosis. You can also order specimens and propagation supplies from a biological supply company. The time spent in larval and pupal stages varies from one species to another, but you may also enjoy watching the larvae go through several instars before they pupate. Be sure to release the adults into the appropriate habitat soon after they emerge.

Monarch butterfly caterpillars are easy to obtain, because they are one of the few creatures likely to be found eating milkweed leaves. In fact, you can plan ahead by collecting and planting some milkweed seeds in clay pots and leaving them outside over winter. In spring you will have fresh milkweed plants that you need only enclose in a cylinder made of clear plastic or wire screen with a screened lid. Embed the bottom of the cylinder in the soil of the clay pot, put in your caterpillar, and watch the show!

Above The stridently-colored, bright warning stripes of the caterpillar of the frangipani hawk-moth (*Pseudosphinx tetrio*), from Brazil, defy predators.

nature of the newly-hatched offspring, most species lay a large number of eggs.

Asexual reproduction

Among some insects there exists a type of asexual reproduction – *parthenogenesis* – in which an unfertilized egg will develop into an adult. This is particularly common among social bees and wasps, where the division of labor is drawn strictly along sexual lines. *Workers*, who perform all of the tasks (apart from reproduction) necessary to keep the hive operating, are all sterile females that develop from fertilized eggs, while males, whose sole function is to fertilize the queen, develop from unfertilized eggs.

Below Workers of the yellow meadow ant (*Lasius flavus*) arrange different forms of cocoon in their galleries. The powerful jaws of the workers grasp the loads, searching out a good grip.

What you can do

Egg hunt

You needn't wait until Easter to have an enjoyable egg hunt. Insect eggs are abundant in spring and summer, but even in winter you may find eggs attached to twigs, branches, trunks, or fence posts, or hidden away in cracks. During the warmer months, the eggs of plant eaters are easily found on the surfaces of leaves. Contrary to popular belief, the eggs of many insect species are large enough to be seen if you look in the right places, although a hand lens is always useful for observation. However, if you're expecting typical "egg-shaped" eggs, like those of chickens, you'll be surprised. Insect eggs come in all manner of wonderfully sculpted shapes. Some look like golf balls, others like milk bottles, barrels, spheres, light bulbs, bullets, or flower pots, to name only a few shapes. Eggs may be laid singly, in rows, in clusters, or on top of one another. Those of mantids are encased in a frothy foam that hardens into a tough protective case, and some moths emb ed theirs in hairs plucked from their abdomens. In addition to the myriad of shapes, insect eggs also come in every conceivable color. Within an individual species, however, all eggs are identical.

Butterfly eggs

 Monarch

 Silver-bordered fritillary

Falcate orangetip

 Tiger swallowtail

INSECT SENSES

We humans tend to imagine that all creatures perceive the world in the same way we do, but such is not the case. Many animals see no colors, while others can see colors we cannot, and vice versa. Some can only detect degrees of light but no images, and still others are totally blind. There are sounds well above or below our range of hearing that are perfectly audible to other organisms, and our sense of smell is notoriously poor compared to that of the so-called "lower" animals. There is much more going on than we realize, and insects know this.

Take color, for instance. Did you know that ultra-violet is a color? We hear about it, and we know that it exists, yet we cannot see it; insects can, however, and they respond to it – a fact not lost on those flowering plants that depend on insects for pollination. Red, on the other hand, a warm and attractive color to us, is invisible to insects.

Eyes

Even the colors mutually visible both to insects and ourselves do not translate into similar images in our respective brains. The head of many mature insects has from one to three simple eyes, or *ocelli*, that serve only to detect various degrees of light. The two *compound eyes* are composed of anything from a few to several thousand individual units called ommatidia, each one evidenced by a separate lens, also known as a facet. Beneath each facet is a second, conical lens, and the two work together to focus light down a light-sensitive structure, the *rhabdome*; this is connected to the optic nerve, which leads to the brain. The quality of the images they generate is not known, but they are supremely adapted to detect motion. Since each unit of a compound eye is stimulated separately, every motion is multiplied many times, and the resulting effect must be rather like watching the same channel on hundreds of television sets at once.

Taste, smell, and touch

The world of insects consists more of patterns of smells and tastes than of light and sound. Taste, smell, touch, and sometimes hearing, are all functions of minute, hairlike bristles called *setae*,

Below A close look at the face of a powerful dragonfly predator (*Aeschna cyanea*) reveals the enormous compound eyes, which meet over the top of the head, giving superb 300° vision.

What you can do

Flower power

Equipment:
assorted photographic
filters
about 8½ x 11 in of white
cardboard
masking tape
flower press
field notebook
pencil
folding table
aerial net
examining jar or bug box

What attracts insects to
flowers – scent or color?
The answer may be either,
depending upon the
species of insect in
question. You can carry
out a simple experiment in
your flower garden or in a
meadow to determine
some of the insects that
respond to scent and to
various colors.

1 You will need to
purchase plastic or glass
photographic filters that
transmit only one color of
the spectrum. If possible,
get 7 filters, transmitting
only red, yellow, blue,
orange, green, violet, and
ultra-violet light
respectively. You will also
need a clear filter as a
control. In the garden or
meadow, observe which
flowers seem to be
popular with flying insects.
Cut eight blooms of a
single species (provided
they are not rare or
endangered) and press

them in a large book
overnight, using tissues or
construction paper to blot
any moisture. Mount each
of the pressed flowers right
side up between a filter
and white cardboard, and
seal the edges with tape.

2 On a sunny day, lay out
the pressed flowers on a
folding table in the
meadow or garden, and
observe which ones draw
more insects. You may use
an aerial net to capture
individuals for
identification and record
them, noting also the type
of flower and time of day,
in your field notebook if
you wish.

3 More than likely, you will
observe more visitors at

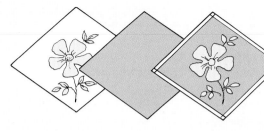

white cardboard with
pressed flower

photographic
filter

filter taped over
cardboard

the ultra-violet and clear
filters than at the others, in
which case you may
deduce that this particular
flower utilizes ultra-violet
to attract insects. Insects
that visit other flowers, but
remain indifferent to the
pressed blooms, are
probably attracted by
scent and not color.
Likewise, flowers that

attract few visitors when
pressed, but many while
alive, undoubtedly rely on
their fragrance to invite
pollinators. It should prove
interesting to repeat this
experiment several times,
using a different species of
flower on each occasion,
and also on cloudy days,
and compare the results.

which may occur all over the body but are usually concentrated on the antennae, mouthparts, and legs. Among the *chemosensory* setae, those detecting airborne chemicals account for smell, while those that perceive chemicals through direct contact with solids or liquids allow an insect to taste. *Tactile* setae, which detect touch, are connected to the cuticle by a ball-and-socket joint, the slightest movement of which stimulates nerve endings on the underside of the joint.

Although insects have no noses, their olfactory sense is keener than anything we can imagine. They smell primarily with their antennae, paired appendages located between the compound eyes and above the mouthparts. Segmented, flexible, and densely covered with microscopic setae, antennae have many configurations throughout the insect world. Among their numerous functions, they serve to detect *pheromones* (chemical messages) emitted to communicate with other members of the species, and to locate food, water, and suitable sites for laying eggs. Depend-

ing upon the insect, antennae may also be used to perceive touch, taste, and/or sounds. Insects also taste through the chemosensory setae on their mouthparts and legs.

Sensitivity to sound varies widely among insects, as do the anatomy and location of their auditory organs. *Phonoreceptors*, as they are called, may simply be modified setae on the body, appendages, or antennae that can detect air vibrations, or they may be more complex structures called *tympanic organs*, located on the abdomen, thorax, or the forelegs. A tympanic organ consists of an exposed *tympanic membrane*, covering an underlying air sac. Auditory nerves are connected either to the air sac or directly to the membrane, which vibrates in response to sound waves, stimulating the nerves in a fashion very similar to the human ear. Some insects can differentiate pitch, or various frequencies of sound, while others are tone deaf. As a group, they can detect sound over a much broader range of frequencies than can people.

Right Insects have one pair of segmented, flexible antennae that are densely covered with microscopic setae. Depending on the insects, antennae may be used to taste, touch, smell, detect pheromones, sound, food, water and sites for egg-laying.

Below This exploded view of the microscopic setae which cover the branches of the antennae shows how messages are passed to the central nervous system. The setae are highly sensitive, functioning primarily as organs for taste, touch and smell.

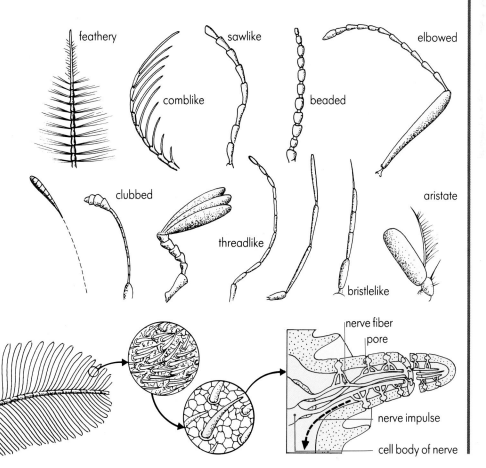

Above Insect antennae are marvellous sensory detectors. The antennae of males may be more branched than those of females, as in this Indian moon moth (*Actias luna*).

INSECT BEHAVIOR

Behavior is the response of an animal to stimuli from its surroundings. A stimulus is detected through one or more senses and interpreted by the brain, which then instructs the body to react, either on the basis of past experience (learned behavior) or in a genetically pre-determined course of action (instinctive behavior).

Understanding an animal's behavior begins with understanding the motive of its genes, those molecular blueprints of all life forms. Genes are very selfish. They care nothing for the rest of their species, only that their host remains healthy and strong and survives to produce as many offspring as possible, thereby passing on the maximum number of its genes to future generations and ensuring the genes' survival. Even behavior seemingly unrelated to reproduction is ultimately oriented toward keeping the animal fit and helping it survive to sexual maturity.

Insect behavior is largely instinctive. Moreover, the behavior pattern of an insect is practically identical to that of every other member of its species, varying only through random jumblings of a portion of the genetic code, better known as mutations. Mutations occur regularly in all populations, resulting in variations in physical appearance or behavior. Some are beneficial, some detrimental, but most have little effect on an individual's fitness. They are the mechanism of evolution through natural selection.

Mutations

The tremendous success of insects is due to both their stereotyped behavior patterns and relatively frequent mutations. Fixed behavior patterns, guiding insects to react in the ways that best promote their survival and reproductive success, enable them to flourish in their preferred *niche*, that portion of an ecosystem that fulfills their needs. However, the world changes, and the inability to change with it is the chief cause of extinction. Because of the fairly short life cycles and great reproductivity of insects, it is likely that when change occurs there will be mutated individuals floating around within a species that are able to cope better than others, meaning that more of their "altered" genes are passed on. With a high rate of reproduction, and decreased competition

Above This diligent worker honeybee (*Apis mellifera*) is collecting pollen grains from the stamens of a spring crocus, and will carry its load away in the pollen-baskets on its hind legs.

from those that could not cope, it is not long before the beneficial mutant gene displaces its predecessor and the species has evolved to meet the new challenge.

Behavior patterns

It would be impossible to list all of the specific behavior patterns unique to individual insect species. As a practical entomologist, you are best advised to learn the basic behavior patterns of the animal kingdom and to try to recognize these among the insects you observe. Significant behaviors will be discussed later in individual chapters on insect orders. The following are some common categories of behavior.

Courtship is ritualized behavior, employed to attract a mate, and is usually actively engaged in by males. The female response, also highly stereotyped, indicates acceptance or rejection.

Copulation, the physical joining of members of the opposite sex, culminates with the transfer of sperm from the male to the female.

Egg-laying is an important clue to the life cycle of an insect, because individual species are quite stringent in their site requirements. Eggs are nearly always laid either on or with easy access to a food source suitable for the larvae or nymphs.

Feeding may encompass predation, grazing, or scavenging. Social insects actively gather food

and return with it to their colony to store it or to share with others.

Defense may be either active, as in insects that sting, bite, or emit noxious chemicals, or it may be passive, employed by those that hide in crevices and under stones, or it can take the form of camouflage in the case of species that "hide" out in the open on a matching background.

Communication, the deliberate transfer of information, occurs essentially between members of the same species. Insects may communicate via all five senses known to us, and we must not discount the possibility of their communicating through unknown methods as well.

Grooming is just as important to insects as to the furry animals to which it is normally attributed. Insects' antennae, eyes, wings, legs, mouthparts, and the hairlike sensory setae covering their bodies must all be kept clean in order to function efficiently.

What you can do

Insect semaphores

Equipment:
penlight
aerial net
examining jar or bug box

One of the favorite summer activities of rural children in temperate regions is to catch fireflies outside on a balmy evening. You can take this fun a step further by learning the language and "talking" with fireflies.

Firefly beetles have luminous chemicals in their tip that can be turned on and off as easily as we operate a light switch. Exactly how they do this is still unknown, but the significance is clear: each species has its own pattern of flashes, prefers specific habitats, and flashes only during certain periods of the night, all of which help members of each species to identify one another. There are many ways in which signals may vary: the number of flashes in a signal, the duration of the signal, the time interval between signals, the distance flown between signals, and even the color of the flashes.

While emitting synchronized flashes, male fireflies fly over an area likely to contain females. When the wingless, larvalike female (also known as a glowworm) recognizes the correct sequence of her species, she signals back. As they continue to signal each other, the male quickly closes in, lands, and copulates with her.

After observing flash patterns for a time, you may be able to duplicate them well enough to elicit responses from the insects themselves. Using a penlight with a sensitive switch, you can repeat the male's pattern downward from about chest height or the female's response upward from the ground. You might even capture one for identification and record the species and its flash sequence in your field notebook.

Above Larva of the *Lamprigera* sp. glowworm from the rainforests of Sumatra. Glowworms are predatory insects which feed on other invertebrates. The females glow to attract males.

33

SOCIAL INSECTS

By and large, most insects are solitary, indifferent to the company of others of their species except during mating and chance encounters. There are quite a few exceptions, however, and these have been lumped into one group called social insects. Termites, ants, most bees, and some wasps are true social insects. Many others will congregate for one reason or another at some point in their lives, but they do not stay together in organized societies throughout their life cycle.

Social insects live in cooperative, interdependent colonies of a single species, usually with a sharp division of labor among social *castes*. Such insects regularly care for the eggs, larvae, and pupae in the colony, a characteristic rare in the rest of the Class Insecta. They build nests of various degrees of complexity, and usually they have all descended from one female, the queen. They also practice *trophallaxis*, the mutual exchange of food and other desirable substances between members of the colony.

The evolution of social life among insects can be divided into several stages. Most insects do not live long enough to see their offspring, making parental care difficult. Probably the first significant step toward a social life occurred when solitary wasps and bees began to build nests, which they provisioned with food and in which they laid eggs. This practice, which still occurs among many solitary wasps and bees, represents a basic parent-offspring relationship, even though the two generations never see each other.

As their lifespans grew longer and began to overlap those of their offspring, the next step, also evident among some modern species, became possible. The mother lays her eggs in a nest with few or no provisions, but returns with prey periodically and feeds the larvae directly. Trophallaxis probably co-evolved as an inducement to this behavior, since the larvae of many social insects, when fed, are stimulated to secrete substances relished by the adults, providing the latter with a powerful incentive to provide food.

The beginning of a social structure is illustrated by certain solitary bees, which we call *sociable* because of their tendency to build their single- or multi-celled nests adjoining those of others of the same species. From this point, it is an easy

Above Worker termites (*Termites longipeditermes*) from Penang, Malaysia, carry in their mandibles balls of powdered lichen, collected from branches for their galleries. Collected vegetable matter is injected with a fungus, which the termites harvest when it has grown.

Above Creatures of habit with a strong group instinct, these driver ants (*Dorylus nigricans*) from Kenya have carved an ant highway with their continued traffic. Like other social insects the ants communicate in part with volatile pheromones, which elicit rapid responses when dangers threaten.

Left Worker ants of *Myrimica ruginodis* tend their larvae in a formicarium. One of the workers' main functions is to care for the eggs, larvae and pupae of the ant colony.

35

transition to a division of labor. Individual bees may adopt specialized tasks, such as guarding the entrance to the nest. While such action obviously benefits the insect's own offspring, those of the other bees also profit.

Caring for the offspring of another seems to violate the theory of the selfish gene discussed under "Insect Behavior," which asserts that the ultimate goal of (genetically programmed) behavior must be the survival of one's own genes. However, since all members of the colony descended from one queen, her offspring are basically siblings to all others in that society. Caring for them ensures the survival of any individual's genes, even if that individual did not contribute those genes directly. Of course, the bees themselves are not conscious of all this. They simply respond as their genes direct them.

Social castes

There are really only two major castes among social insects: the reproductive caste, composed of males and queens, which has the sole responsibility of making more offspring, and the nonreproductive caste, consisting of only workers or workers and soldiers, which undertakes all of the work in maintaining the colony, including building and repairing the nest, collecting food, and caring for the eggs, larvae, and pupae.

Social bees and wasps have evolved an effective means of controlling the numbers in each caste to ensure that the colony operates efficiently. Only a limited number of males are needed to promote competition for the privilege of mating with the single queen. Extra queens leave with some males to start their own colonies. The queen has control over whether a given egg gets fertilized; unfertilized eggs develop into males, and fertilized ones become females. The queen, in turn, is stimulated to either fertilize eggs from sperm stored in her abdomen or withhold sperm based on the sizes of the cells built by workers. Into smaller cells she deposits fertilized eggs, while larger cells receive unfertilized eggs. Nutrition determines if a female grows into a worker, all of which are sterile females, or a queen. Among wasps, underfed or improperly fed females become workers, while well-nourished females grow into queens. Honeybee workers secrete a white paste, called royal jelly, from glands connected to their mouthparts and feed this to all larvae for at least their first three days. Those females that receive royal jelly throughout their larval stage develop into queens, and the rest become workers.

Right Social wasps (*Polybia occidentalis cinctus*) from Trinidad build a nest below a protective leaf in the rainforest. The nest is made entirely from wood fiber collected from tree trunks and chewed to make combs. The tiers of a wasp comb lie horizontally, compared to those of bees, which are vertical, and wasps remain on the surface of the comb to attack predators. The nest also has an outer protective covering.

Round dance

Waggle dance

90°

What you can do

Make friends with a beekeeper

One of the best ways to learn the intricacies of social insects is to take up beekeeping as a hobby. Unfortunately, few of us have the time or finances to devote to this pastime, so the next best thing is to make the acquaintance of a beekeeper. Ask around at the farms, orchards, nature centers, or health food stores in your area and you will probably find out the names of several potential contacts.

Don't be bashful; most people love to share their hobbies or occupations with interested individuals, especially if they are good at it. They may even have extra protective equipment and invite you on their rounds. Do a little homework beforehand so that you are familiar with the terms they use.

Above Hands-on experience with honeybees, in the presence of a beekeeper, allows children to see brood, honey and pollen stores.

Above Worker honeybees communicate information about pollen and nectar sources to members of their hive by "dancing." The locations of food sources within 80ft (25m) of the hive are communicated by a round dance. More distant food sources are indicated by a waggle or figure-of-eight dance, which involves waggling the abdomen from side to side during the straight portion of the figure-of-eight. Distance is indicated by the length of the run and the number of waggles. The angle of the straight run corresponds to the angle between the direction of the food source and the sun as viewed from the hive. The waggles and buzzes may indicate food quality.

OFFENSE AND DEFENSE

The variety of the insect world has given rise to a great number of protection mechanisms. First and foremost is the insect's durable external skeleton, a veritable suit of armor for many species. Not only does it offer protection against physical forces, but it also prevents undesirable water loss and is resistant to a great many chemicals. In addition, the physics of having muscles *inside* the skeleton gives insects tremendous strength in relation to their small size.

Size itself is a great defensive asset of insects. Being small enables them to go about their lives unnoticed by many larger animals that would eat them. In fact, their diminutive stature is a major reason why entomologists estimate that there may actually be twice as many insect species or more on earth as those already classified – we may have simply overlooked the rest so far. Many just take refuge in a crack or under leaves or stones until danger has passed, and quite a few spend their entire lives out of sight of most vertebrates.

Of those that avoid danger by hiding, a great many rely upon some form of camouflage to enable them to hide out in the open. Some incorporate colors and patterns on their exoskeleton or wings to match the background material upon which they normally rest. Others have evolved fantastically modified body shapes and appendages that perfectly mimic plant parts such as twigs or leaves, right down to the swaying motion of a leaf in the breeze. Notable among these are the stick insects (Order Phasmatodea), leaflike katydids (Order Orthoptera), and treehoppers and leafhoppers (Order Homoptera).

Many of those insects that cannot hide so easily have another option – flight. As a group, winged insects have complete mastery of the air and can execute maneuvers beyond the capability of any other creature or machine. Of course, some are better fliers than others, but even the worst are better off than they would be if they could not fly at all.

Self-advertisers and mimics

In contrast to those which remain inconspicuous, a great number of insects boldly advertise their presence with bright colors or contrasting patterns. Most of those so marked have an unpleasant experience to offer potential predators, such as a

Right A female emperor moth (*Saturnia pavonia*) waits patiently on a grass stem, just after emerging from her cocoon. Her silvery colors, made up entirely of scales, portray a pair of fearsome "false-eyes" to deter predators. If the moth is disturbed, she will show off a further pair on her hind wings. She also gives off a powerful pheromone while waiting for males to gather round and compete for the chance to mate with her.

Left It is often the case that the smaller the moth, the more intricate the pattern. The coloration of this small moth, *Alabonia geoffrella*, has evolved as camouflage against predatory species.

venomous sting or spines, irritating hairs, repellent glands, or toxic or distasteful compounds in their tissues. Once so educated, an animal is unlikely to make the same mistake twice, and usually gives that species a wide berth. One might think that traits inviting attack would be quickly eliminated from the population, but apparently many more such insects are spared from repeat attacks than are lost to ignorant predators.

Riding on the coattails of these conspicuous insects are mimics, whose chance resemblance to a species habitually avoided by predators has encouraged their evolution to the point where it requires very close scrutiny to tell the harmless species from the dangerous one. Classic examples of this are the monarch and viceroy butterflies of North America. As a larva, the monarch, whose hazardous migration of thousands of miles each autumn has endeared it to millions of Americans, feeds exclusively on milkweed plants. Milkweeds produce a milky latex containing cardiac glycosides, toxic substances that cause nausea and vomiting in small doses and death to vertebrates in large doses. These toxins deter nearly all animals from grazing upon milkweeds, but monarchs have adapted to not only feed on them but to incorporate the cardiac glycosides throughout their tissues, making them extremely distasteful to predators, especially to birds that might otherwise pick them out of the air. Orange-and-black viceroy butterflies are a product of the same natural selection processes that govern monarchs; those that, through mutations, more closely resemble monarchs will more successfully pass on those traits to future generations.

Defense by offense

Social insects have adopted the axiom that the best defense is a good offense. While many insects are capable of delivering a painful bite or a venomous sting, the true power of this defense is unleashed when it is multiplied by hundreds or thousands of members in a colony. Social insects communicate extensively via pheromones, chemical messengers that elicit immediate instinctive responses. Alarm pheromones are potent and travel rapidly through a colony, rallying its occupants to repel any intruders.

What you can do

Gross grasshoppers

Equipment:
sweep net
bug box
hand lens

Many insects, when captured, give off a repulsive fluid to persuade their enemies to release them. The "tobacco juice" that a grasshopper spits as it is handled is really the foul-tasting contents of its gut. This is often effective in securing the insect's release, giving it just enough time to escape.

In summer, sweep an area of tall grass with your net until you catch a grasshopper. Holding it securely by both femurs will immobilize it for observation. Gentle prodding may encourage it to release a brown fluid from its mouth if it hasn't already. What does it smell like? Handle with care, as this substance will stain skin and clothing.

ENEMIES AND ALLIES

Insects have not had a fair deal. They are not popular subjects of nature study, in part because we tend to associate them with nuisance and with the transmission of diseases. As the most numerous and diverse class of organisms on earth, they are inextricably woven into the web of life, and so there are inevitably some adversarial relationships with humans. However, we also have a good many insect allies.

Beneficial insects

For one thing, insects are responsible for perpetuating many of the green plants that feed our planet and oxygenate its atmosphere. With the evolution of flying insects, plants were quick to take advantage of these reliable pollinators. Insect-pollinated plants have incorporated colors, fragrances, and nectar attractive to various insect species into their flowers, and many have even adapted to open their flowers only during the times of day when their primary pollinators are most active. Bees, butterflies, moths, and flies are the major pollinators. Hives of honeybees, perhaps the world's premier insect pollinators, are therefore frequently kept in orchards or nurseries, and on farms. As a side benefit to their pollinating activities, the bees supply us with large quantities of honey and beeswax, which is used in lubricants, salves, ointments, furniture polish, candles, and other products.

Other insects also benefit plants and, indirectly, us. The activity of burrowing insects, such as ants, loosens, aerates, and fertilizes the soil, encouraging the development of healthy plant roots. Some species feed upon noxious weeds, helping to control them, and predatory insects will readily take any insect within their power, including those damaging to plants. The value of insect predators, such as ladybird beetles, mantids, and lacewings, is legendary.

Insects have supplied us with a good many other products besides honey and beeswax. Most of the world's silk, for instance, is produced from the cocoons of the silkworm moth, *Bombyx mori*. Sometimes the insects themselves become the product; among certain cultures, insects, such as grasshoppers, flying ants and honey ants, are popular fare.

Harmful insects

Despite the fact that most insects have no direct interaction with humans at all, there are some that are unquestionably harmful. Mosquitoes can transmit such diseases as malaria, yellow fever, encephalitis, and elephantiasis. Fleas are the vectors of bubonic plague, typhus, and tapeworms. Houseflies have been implicated in the spread of tuberculosis, typhoid, cholera, amoebic dysentery, anthrax, and other illnesses. Human lice, though they are basically parasites that usually cause only discomfort, also may transmit typhus and relapsing fever. One of the best known disease vectors is the tsetse fly, which transmits African sleeping sickness.

Ever since the dawn of agriculture, people have done battle with insects over the crops that both

Above Silverfish are viewed as pests because they damage books and papers. This member of the silverfish family (*Lepisma saccharina*) reveals the silvery color responsible for its name.

Above The gardener's friend – ladybird beetles feeding on aphids.

Left A small swarm of "killer" bees hangs patiently in the branches waiting for "scout workers" to find suitable accommodations. "Killer" bees are a hybrid, the result of breeding between stocks of African bees and Latin American bees (*Apis mellifera*). Their aggressive nature is notorious.

desire; so far, the conflict seems to be a draw. The use of chemical pesticides during the 20th century has given humans the edge, but their use must be carefully regulated to avoid worldwide environmental disaster.

Insects may destroy crops in one of two ways. Most obviously, their consumption of the plant means there is less for us. Grasshoppers, true bugs, aphids, and beetles are the most serious offenders in this category. Infestation of our stored grains and cereals is also a serious problem, but most of this damage is caused by only a few species of beetle. Perhaps more serious are the many plant diseases introduced through the wounds where insects have been feeding, boring, or laying eggs.

Adding insult to injury, certain insects inflict considerable damage on our possessions as well.

Termites are a worldwide threat to any wooden structure. Carpenter ants and wood-boring beetles pose a similar but much lesser problem. Silverfish (Thysanura) and book lice (Psocoptera) can destroy books and papers as they feed on paste, glue, and the sizing used to create glossy pages. Clothes moths were once significant destructive pests of natural fabrics, but their significance diminished with the increased use of synthetic fibers. Carpet beetles (Family Dermestidae) are a much more serious problem today, as they attack both natural and synthetic fibers.

Despite the serious nature of insect pests, it behooves the practical entomologist to dispense with any prejudice that may interfere with the study of this fascinating field of natural history. Bear in mind the many beneficial insects, and the fact that most insect species are basically benign.

ENTOMOLOGY IN ACTION

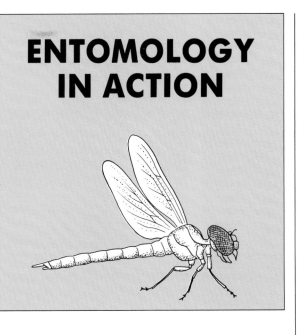

TOOLS OF ENTOMOLOGY

The beauty of entomology as a pastime is that you can find insects almost anywhere and at any time. You can spend a few minutes here and there or a whole day, depending upon your level of interest and time available, and with such a wide array of subjects, you need never be bored. Best of all, you need very little specialized equipment to get started. Even so, a few appropriate tools will yield much more for your efforts.

Tools for examining specimens

Since adult insects rarely stand still, you will need to confine live specimens in some way in order to study their features. Any wide-mouthed clear glass or plastic jar will do, but the smaller ones are generally more suitable. Especially useful are bug boxes, which are clear plastic cubes with a magnifying lens in the lid. These come in a variety of sizes and can often be purchased in the gift shops of natural history museums and nature centers. If your subject is particularly active, you can slow it down by placing it in the freezer for a few minutes.

A good quality glass hand lens, available from biological supply companies, will reveal minute details of insect anatomy that you may otherwise overlook. A magnification of 8x to 10x is optimal, and some models have several glass elements that can be used in different combinations to vary their magnifying power. Photographic loupes sold in photography shops will also serve well.

Despite the disclaimer that practical entomologists do not need expensive equipment, the author does recommend one such instrument for those with a serious interest in natural history. A stereomicroscope, also called a dissecting or binocular microscope, is a wonderful tool that provides an unequaled three-dimensional view of small objects. It will usually have a variable magnification from 10x to 60x, and is invaluable for viewing or dissecting insects. Several models are available through biological supply companies, and while they are not cheap, most are affordable, and nearly all cost less than good-quality conventional light microscopes.

Examining insects against a contrasting background will often help you to see their finer features. Keep small pieces of black and white cardboard, about four inches square, with your equipment for this purpose. A small penlight will be useful under low light conditions.

Tools for handling insects

There are several implements that are handy for manipulating captured insects. Forceps – similar to tweezers but with a more sensitive touch that is less likely to damage your specimens – are useful in handling specimens. Probes may be used to prod an insect into moving in a certain direction or to spread the wings and legs of killed insects for examination. Both probes and forceps are easy to make and are also available from biological supply companies. To move small killed insects, a camel hair artist's brush is ideal.

Tools for dissecting insects

You will find a dissecting tray handy for examining or dissecting killed insects. You can easily make your own by pouring melted paraffin into a small baking tray to form a layer of wax about one centimeter deep. When the paraffin becomes too full of holes or gouges, simply melt it again and let it solidify. Pins of assorted sizes are useful for securing insects during dissection, and can be ordered from biological supply companies.

Should you wish to dissect any specimens, you will need cutting instruments. Scalpels are available commercially, but expensive. A good substitute is an X-acto razor knife with interchangeable

blades, available in hobby stores. An old travel toothbrush holder makes a safe carrier for it. A fine pair of scissors may also prove useful.

Field guides

Invaluable in the study of most other fields of natural history, field guides are of more limited worth in entomology, primarily because no single guide can list even half of all insect species in a given region. Even so, the insect field guides that are available usually list at least the most common species in the geographical areas they cover, and so deserve a place in your library.

Above There are three main reasons for capturing insects: to aid in identification, to study their behavior and life cycles, and to create a collection. There are a number of methods you can use, but in all cases the tools required are comparatively simple, ranging from nets to beating trays and traps.

All good field guides have several features in common. They should be logically organized, preferably by order and family rather than by some nebulous criterion, such as color, size, or habitat. Detailed illustrations are a must, as are range

maps or descriptions, specific habitat preferences, and clear physical descriptions of each insect, including characteristic identifying features. Some also list interesting facts about each individual species, such as its life cycle, behavior, and significance to humans.

Identification keys

Useful in conjunction with field guides are identification keys, sometimes called dichotomous keys, that guide you with certainty to the order and family of your specimen. At each step, the key will present two options, only one of which will apply to the insect in question. Whichever one you select will either reveal the order or family of your specimen or guide you to the next step in the key. Keys to insect orders and families are not as common as field guides, but you may find them in certain field guides, in trade books sold in book stores, or in entomology textbooks at your local library. An entomology professor in any university may also be willing to make copies of keys for you.

Keys to individual insect species would be cumbersome, and the features separating one species from another are often so minute that they are impractical to discern. Fortunately, the nature of entomology is such that knowing the order and family of an insect will provide plenty of clues about it.

Recording what you see

In many respects, the most useful tool of any practical entomologist is a well-kept field notebook. Knowing the name of a species is no prerequisite for making observations of its appearance and behavior. You can even make up your own name for a specimen as you take notes, thus giving yourself a handle on that species until you can make a positive identification.

Look in stationery stores to find a medium-sized notebook, strongly bound and composed of high-quality, unlined paper that will not tear easily. Keep it as clean as possible to preserve its usefulness. Self-sealing food storage bags make tough, waterproof storage cases for your notebook, field guides, keys, and maps while in the field.

Keep a pen and several pencils in your day pack along with the rest of these tools. Backpacking and outdoor shops sell space-age pens that write in any position and under any conditions. A short ruler is also handy for recording dimensions.

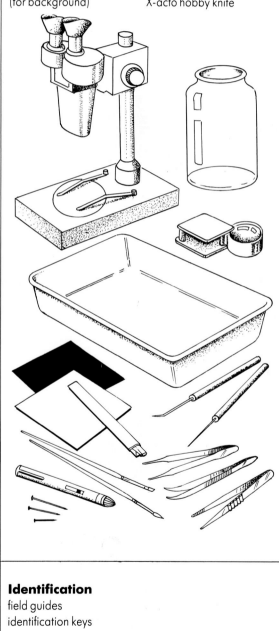

Examination
hand lens
(stereomicroscope optional)
clear jar or bug box
penlight
square of black cardboard (for background)
square of white cardboard (for background)
dissecting tray
pins of various sizes
camel hair artist's brush
probe
forceps
X-acto hobby knife

Identification
field guides
identification keys

Tools of the practical entomologist

Recording
field notebook
pens and pencils
ruler
camera and accessories

Collecting
aerial net
sweep net
dip net or kitchen strainer
small white plastic or metal
tray
aspirator
killing jar

specimen box
relaxing chamber
triangular paper
envelopes
insect pins of assorted
sizes
pinning block
spreading board

display boxes
paper points and labels

Miscellaneous
day pack
topographic maps of local
area
compass
Swiss army knife
water bottle

CAPTURING INSECTS

There are many methods of capturing insects, and any one is acceptable as long as it is safe and does not damage the insect. Many insects are easily collected by hand, with forceps, or by simply clapping a jar over them. Others require a bit more finesse and some basic tools. The following are a few such devices that are either simple to make or else inexpensive to buy.

Nets

Many insects are adept at evading or escaping predators and can easily evade our clumsy efforts. Others can be dangerous to handle; bees and wasps inject venom as they sting, and many of the larger true bugs can deliver a painful bite. For these, and for insects that hide in tall grass or underwater, we need a net, of which there are three basic types.

An aerial net, with which most of us are familiar, is used for netting flying insects out of the air. It is

Above The choice of a suitable habitat for an invertebrate or an insect depends on knowing the ecological requirements of the species, and spending time in the field helps one to understand various groups. It is essential to be prepared and to take along the right equipment.

a long sack of soft nylon, organdy, or silk netting that will not normally damage specimens. This delicate material does not wear well and should not be used to capture insects in vegetation or on the ground. After netting an insect with an aerial net, give the handle a quick half-twist to double the net over the opening and prevent escape. The best way to retrieve your catches is by inserting a jar into the net and trapping the insect between the netting and jar.

Sweep nets are made of heavy white muslin or canvas and are used with a sweeping motion to capture unseen insects in tall grass or weeds. This "shotgun technique" rarely fails to turn up something of interest. It is especially useful in meadows during the summer and fall.

A dip net is a long-handled tool used to capture aquatic insects. It has a flat side that may be placed against the bottom of a stream or pond. The triangular hoop is threaded through a tube of heavy muslin with a screen of heavy nylon netting on the end, to allow water to flow through. Insects may be difficult to see among aquatic debris, so you may need to dump the contents into a white tray and watch for movement.

Beating tray
Insects that frequent plants are often reluctant to fly or run, preferring instead to remain still and rely on their natural camouflage and small size to escape detection. A sweep net, good for grassy meadows, does not work as well on trees and shrubs; here you need a beating tray. This is a simple affair made of white cloth stretched over a frame of two crossed sticks. (A light-colored umbrella also works well.) It is held or placed under a shrub or tree branch, to which you deliver several sharp blows, taking care not to damage the plant. Insects having taken refuge there will fall onto the beating tray and can then be collected by hand or other means.

Pitfall traps
Pitfall traps are effective in catching insects that scurry across the ground rather than fly. Simply dig a small hole with a garden trowel and sink a jar or cup in it so that the top is flush with the surface of the ground, filling in with soil around the outside of the container. An inch of 30 per cent water to 70 per cent alcohol mixture in the bottom will kill the insects if you do not wish any to escape. If you do

not want to kill any insects, use yoghurt pots with drainage holes and check every few hours. Place three or more strips of wood radiating from the rim like the spokes of a wagon wheel; these will guide insects into the trap. Cover the trap with a slab of wood or stone to keep out rain water, and check it daily. You should set several traps in a variety of locations in order to find the best sites in a given area.

Bait traps
One of the major pastimes of insects is searching for food, and many species can be lured with bait. Pitfall traps or small containers suspended from a tree branch make excellent bait traps. Cover them securely with a large-mesh screen with openings about 2 cm square to admit large insects but keep out other scavengers.

A mixture of equal parts of molasses and water makes a fine bait. Add a little yeast and allow the

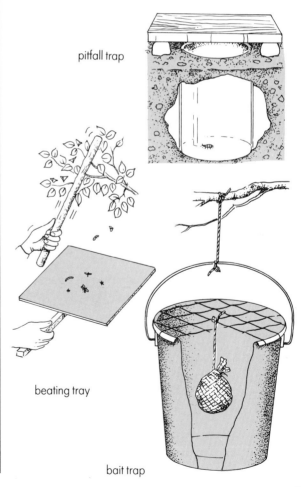

pitfall trap

beating tray

bait trap

mixture to ferment for a day or two before baiting traps with it. A mash of fermenting fruit also works well. Cloth strips soaked in a bait mixture and suspended from a tree branch will attract larger insects, such as nocturnal moths, that cannot fit through the screen covering your other bait traps. The mixture can also be painted onto the bark of a tree. Visit the site every hour or so during the night.

Decaying animal remains or meat scraps will attract many species, especially beetles and flies. If you wish to kill your prisoners, suspend the bait in a cheesecloth sack from the wire screen and put an inch of the water and alcohol mixture in the bottom of the trap. Dung is also a very effective bait, but take precautions against the transmission of disease or parasites by handling it only with disposable implements. The water and alcohol mixture in the trap should sterilize any specimens attracted to the dung.

Berlese funnel

More insects than you might have guessed reside in the cool, moist debris littering a forest floor. A Berlese funnel is a device that can drive insects from such material. Simply place a large funnel, one that narrows to about a two centimeter opening, small-end down in a tall glass jar containing a layer of moistened paper towels. Fill the funnel with leaf litter and suspend a high-wattage light bulb over it. As the debris warms and dries, any insects it harbors will burrow deeper, seeking their preferred conditions, until they reach the funnel opening and fall into the jar.

Sifter

A sifter will do much the same job as a Berlese funnel. To make one, cut the bottom from an old bucket, leaving about a one-inch lip. Cut a circle of large wire screen to rest in the bottom of the bucket. Fill the sifter half-way with debris and cover with canvas, securing the cover with cord or a large rubber band. Shake the sifter vigorously over another bucket or other deep container and collect your specimens.

Light trap

It's common knowledge that nocturnal insects are attracted to light, and you can use this behavior to your advantage. There are many different types of light traps. Perhaps the simplest is a white sheet tied at all four corners and suspended vertically, with a lantern shining on it from behind. As insects alight on the sheet, they may be simply scooped off with an aerial net, allowing you to capture only the ones you want. An empty Berlese funnel with a cylinder of wire screen extending up around the light bulb also makes an effective light trap, though you may need to adapt it by using a funnel with a larger opening at the bottom.

An aquatic light trap can be made with a flashlight, a watertight jar, a length of pipe about six inches in diameter, and some fine-mesh wire

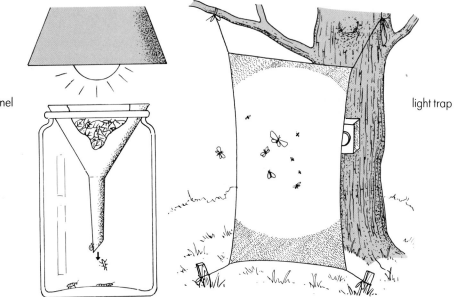

berlese funnel

light trap

screen. Turn on the light; seal it in the jar, and place it in the pipe. Use the wire screen to cover the end of the pipe toward which the light does not point, and fashion a funnel out of the same material for the other end, pointing inward. Secure the pipe with strong cord and lower it into the water, with the funnel facing upstream if there is a current. You may need to weight it down with rocks if the pipe is not heavy.

Aspirator

Some insects that you will find are so small that they cannot be easily collected by hand or with a net. For these, the best method may be the use of an aspirator, with which you manually suck the insects into a small vial.

You can make an aspirator from any small vial with a tight-fitting cap or stopper, two plastic straws, and a length of flexible plastic or rubber tubing. Drill two holes, exactly the same diameter as the straws, in the cap. Insert one straw so that it is about one centimeter above the bottom of the vial. Cut a short piece from the second straw and insert it so that about two centimeters extends into the vial and one centimeter protrudes above the cap. Seal the joints between the cap and straws with silicone bathtub caulk. Cover the inside of the short straw with a small piece of fine netting so that you don't accidentally suck insects into your mouth. Attach one end of the rubber tubing to the other end of the short straw, seal with silicone caulk, and you're ready for action. Turn over stones or sift through forest and plant debris to suck up little insects like springtails.

Top Fogging is a technique now widely employed in rain forest in order to find out which insects live in the canopy. A large gun is hoisted aloft, armed with burning insecticide and a petrochemical, and maneuvered around at first light, disturbing the sleeping insects which then fall to the ground.

Above The rain forest floor is festooned with large canvas funnels for collecting insects. Fogging is carried out when there is little wind so that everything falls into the funnels instead of being driven elsewhere.

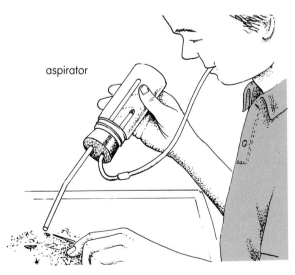

aspirator

KEEPING LIVE INSECTS

Observing the behavior and life cycles of insects provides fascinating insights into their lives, and one of the easiest ways to do this is by watching captive insects. A simple jar with a ventilated lid will serve as a cage for some species, while others may require more upscale accommodation.

If you wish to keep your subjects alive indefinitely, you should identify the species you are dealing with in order to research their needs. At the very least, you will need to determine their orders and families to make an educated guess about their preferred conditions. Like all animals, insects have certain basic requirements: food, water, and protection from extremes in their environment. In order to raise more than one generation, you will also need to provide suitable sites in which they can pupate and lay eggs. Bear in mind that species which overwinter as eggs or pupae may fail to develop if not subjected to low temperatures. You can satisfy this requirement by placing the eggs or pupae of such species in the freezer or outdoors for a few weeks.

Probably the most specific need of your captive insects is food. Water is extracted by some insects, especially plant-eaters, from their food, and most other terrestrial insects thrive on a light misting from a spray bottle, which simulates dew. Be careful, though; large droplets can trap small insects, and too much moisture encourages the growth of mold. As for environmental conditions, the climate in any home is moderate enough for nearly any insect. However, many insects are specialists with precise food requirements. This is particularly true of the herbivores, which often have evolved life cycles centered around a specific plant species that is reliably found in a given habitat. A good example is the monarch butterfly, *Danaus plexippus*, whose larvae eat only milkweed plants. Fortunately, there are plenty of other insects which are easy to satisfy. Scavengers and predatory insects are usually less picky about their foods, since in nature they have evolved to take advantage of whatever comes along.

Terrestrial insect homes

The style of cages you choose for your insect zoo is limited only by your imagination, though they must be ventilated and have lids or doors that

Above Milkweed bugs (*Oncopeltus fasciatus*) mate in characteristic fashion on a leaf. These are from St. Vincent in the West Indies and can walk around in broad daylight in defiance of predators, since their bodies contain poisonous substances gathered from the plants that they feed on; this protection is advertised by their warning colors.

Left Mealworms (*Tenebrio molitor*) are the larvae of darkling beetles, and are often reared in laboratories for study of insects, and as bait for fish. The larvae are pests of cereal, grains, bran, and other stored foods, where they eventually pupate and emerge as winged adults.

close securely. It also helps if they are easy to clean, which should be done periodically. Attempt to simulate natural conditions if you wish your guests to remain alive for a substantial period. Include some soil, sand, or rocks if they are part of the insects' natural niche, and perhaps a stick or plant for them to rest on or climb.

Large, wide-mouthed jars with mesh covers or ventilated lids are simple yet effective cages. If space is not a problem, an aquarium with a snug-fitting screen top also works well. A wire colander from the kitchen makes a good home when turned upside down, provided there are no gaps around the bottom.

A versatile insect cage may be made from a piece of wire screen and two cake pans of equal size. With a tape measure, determine the circumference of one of the cake pans. Cut a rectangular section of wire screen with one dimension equal to this measurement and join the two sides with duct tape or monofilament fishing line to form a cylinder. Rest this inside one cake pan, and use the other for the lid. A pane of glass or plexiglass may substitute for a lid if the screen is sturdy enough to support it.

Insects that live in plant litter can be reared by placing the debris in a closed box with a vial protruding from one side. As adults emerge, they

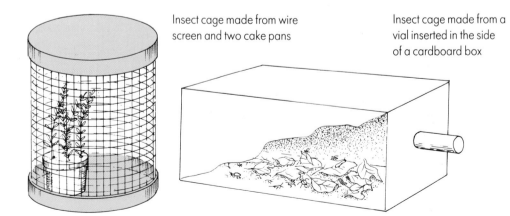

Insect cage made from wire screen and two cake pans

Insect cage made from a vial inserted in the side of a cardboard box

move toward light and will end up in the vial.

Unless you supply plants for food, maintain moderate humidity near terrestrial insects by keeping a piece of moist sponge in each cage. Plants provided for food must be kept fresh or replaced regularly, and potted plants are a better food source than cuttings. If plant cuttings are placed in water, cover the water container, leaving holes large enough for the stems, but small enough to prevent the insects from entering and drowning.

An alternative to simulating natural conditions in your home is to cage herbivorous insects in the field. If you find insects, immature or adult, feeding on a small plant, you can cover the whole plant with a screen cylinder and a lid and visit it periodically to observe them. If they are feeding on a large plant, tie a fine-mesh bag around that portion of the plant.

Aquatic insect homes

Aquatic insects, as one might expect, require an aquarium or a large glass jar. Fill the aquarium with water from the place where you capture the insects, not with tap water. Aerate the water with an air pump if the insects come from a stream; otherwise, this will probably not be necessary. Line the bottom with clean gravel or coarse sand for easy maintenance, and slope it to above water level to provide shallow and deep areas. Most adult aquatic insects can also fly, so a wire screen or glass lid will be needed. Some may need plants or sticks emerging from the water so that they can crawl up and make their final molt into adulthood.

You may wish to create a balanced aquarium by introducing some small native fish, amphibians, other invertebrates, and plants as well, but remember that a crowded aquarium will not stay healthy, and an excess of plants will make the insects harder to see.

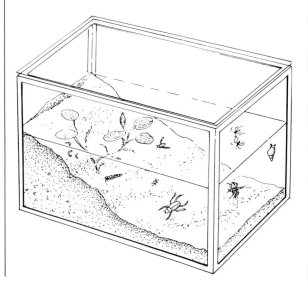

Above An insect aquarium will enable you to observe the life cycle and behavior of aquatic insects, such as dragonflies, caddisflies and mayflies. Clean gravel, plants, and water taken from the capture site will help to simulate suitable natural conditions.

Left This fresh water aquarium could easily be adapted to provide a home for local aquatic insects.

Above This four-spotted chaser (*Libellula quadrimaculata*), photographed in England, has just emerged from its molted exoskeleton, which still hangs on the vegetation. The chaser expands its wings until they are hard and firm, ready for flight. Like that of many newly-hatched insects, its body is clothed in a fine layer of hairs. Note the predator's characteristic large compound eyes with thousands of ommatidia, jaws, and legs suited to catching prey in flight.

MAKING A COLLECTION

Creating an insect collection is not necessary to practical entomology. You can learn a lot by observing the behavior of wild or captive insects, and by examining specimens that you have captured without harming them. Building a worthwhile insect collection is a meticulous and time-consuming endeavor. However, naturalists seem to have a penchant for collecting things, and so a portion of this chapter is devoted to helping you assemble a creditable insect collection.

If you decide to start an insect collection, it should be assembled in such a way as to be of value to the scientific community should you ever decide to donate it to a museum or university. Each collection should be composed of correctly-mounted and well-preserved specimens, and it should also tell a story about where, when, and by whom each specimen was collected. You may choose to specialize and collect only a certain order of insects, such as beetles.

A few words of caution: do not add butterflies or moths to your collection unless you are serious about maintaining it. The Order Lepidoptera is the one major group of insects whose members have shown any serious population declines in the face of human activities. If you do collect butterflies and moths, know the endangered species in your area and refrain from killing or injuring them.

Killing insects

Dispatching insects for your collection requires a killing jar with an airtight lid into which you place the insects, together with a volatile, poisonous substance. Professional entomologists sometimes use potassium cyanide as a killing agent, but this is extremely toxic and very dangerous if not handled with extreme care. By far the better choice is ethyl acetate, which is available from pharmacists and biological supply companies and is far less hazardous. If ethyl acetate is not readily available, acetone (fingernail polish remover) is a good substitute. No matter which of these two you choose, avoid breathing the fumes or spilling any on your skin. Killing jars should be clearly labeled "POISON" and should be kept out of the reach of children.

Killing jars are easy to make. You will need three wide-mouthed jars – large, medium, and small –

with screw-top lids. Pour about a one-half inch layer of plaster of Paris, made with a little extra water so it pours easily, into the bottom of each jar and allow them to dry overnight with the lids off. Tape the bottom third of the jar with electrical tape to reduce the chance of its shattering if dropped. Just prior to using each jar, sprinkle a few milliliters of ethyl acetate onto the plaster of Paris, which will absorb it, and close the lid for about five minutes. You will need to recharge the jar with ethyl acetate every few days, depending on how often it is opened. Reserve the large jar for

butterflies and moths, which tend to lose wing scales that could foul other specimens. Wipe the jars clean periodically.

Transporting specimens
Rarely will you mount specimens in the field, so you will need some means of transporting them without damage. The killing jar can serve this purpose if there are only a few insects. Otherwise, purchase an ordinary small plastic food storage container with a snap-on lid; line this with absorbent cotton, and sandwich your specimens

Above This cabinet with glass-topped drawers displays insects in a traditional manner, enabling comparisons to be made. Familiar to Victorian naturalists, the insect cabinet is now rare.

between thin layers of cotton. Place each butterfly and moth in an individual triangular paper envelope first, in order to protect their wings (see illustration on p45 in "collecting tools"). Insects will dry when stored in this fashion and must subsequently be softened in a relaxing chamber before mounting.

Relaxing chamber

Dry insects become brittle and nearly impossible to mount. Overnight storage in a relaxing chamber, which is an airtight container of high humidity, restores their flexibility, so that legs, wings, antennae, and other parts are less likely to break as the specimen is handled during mounting. Make a relaxing chamber from the same type of airtight container as is used to transport specimens in the field, lining it with moist paper towels or a damp sponge. Rest the insects on a small dish inside to keep them from direct contact with the moisture, and place a separate container inside the chamber with a few milliliters of ethyl acetate or a few grams of moth flakes to inhibit the growth of mold.

Mounting specimens

In order for your collection to be worth anything, the specimens must be mounted properly. In most cases, this entails pinning the insect through the thorax so that it is suspended right side up, and parallel to the floor of the collection case. Special insect pins, available through the biological supply companies listed in the appendix, must be used to mount your specimens. These are available in different sizes, No. 1 being the most slender and No. 5 the thickest that you will need. They are usually black and coated with varnish to make them rustproof. Common sewing pins will damage specimens because they are too thick, and they will rust.

Most insects, including flies, wasps, bees, butterflies, moths, mantids, and cockroaches, should be pinned vertically through the right side of the thorax between the bases of the front wings, with the pin emerging where it will not damage legs or other delicate identifying features on the underside. Beetles and most families of the Order Homoptera are pinned through the front portion of the right wing. True bugs should be pinned through the right side of the scutellum if it is large enough to do so, or through the right wing like beetles if it is not. Grasshoppers, crickets, treehoppers, and leafhoppers are pinned through the right rear portion of the pronotum; grasshoppers may also be pinned with one wing spread on a spreading board. Dragonflies and damselflies may either be pinned vertically, with their wings spread, or horizontally through the thorax, with the left side up and with their wings positioned over

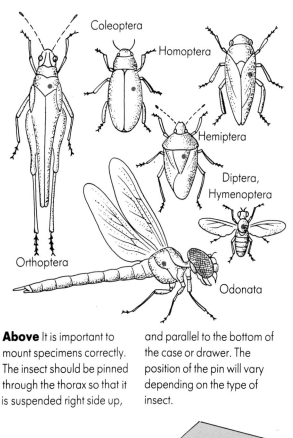

Above It is important to mount specimens correctly. The insect should be pinned through the thorax so that it is suspended right side up, and parallel to the bottom of the case or drawer. The position of the pin will vary depending on the type of insect.

Above A pinning block helps you to pin specimens at a uniform height. Once the pin is through the insect, place the pin in the hole of the 1in (25mm) step and press down until it touches the hard surface under the block. Use the ¾in (18mm) hole for the main information label, and the ½in (12mm) hole for any secondary label.

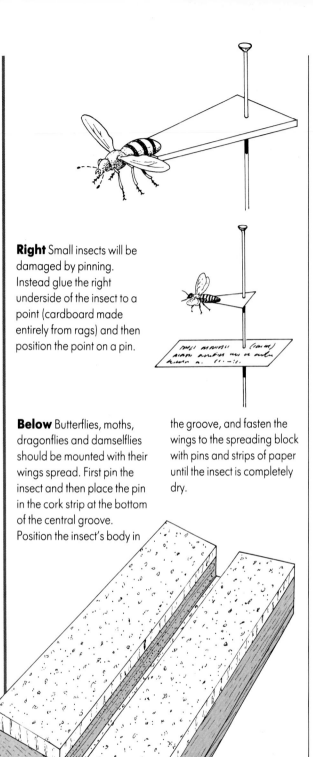

Right Small insects will be damaged by pinning. Instead glue the right underside of the insect to a point (cardboard made entirely from rags) and then position the point on a pin.

Below Butterflies, moths, dragonflies and damselflies should be mounted with their wings spread. First pin the insect and then place the pin in the cork strip at the bottom of the central groove. Position the insect's body in the groove, and fasten the wings to the spreading block with pins and strips of paper until the insect is completely dry.

the back before pinning. Long abdomens, antennae, and other structures that might droop can be supported by pushing the pin deep into a soft surface, such as cork, balsa wood, or styrofoam, positioning the parts as they rest on this surface and leaving them until they dry.

Pinning block

Mount specimens by holding them between the thumb and forefinger of one hand and inserting the pin with the other. All specimens should be pinned at a uniform height to a pinning block, available from biological supply houses. The most useful pinning block has steps 25mm, 18mm and 12mm high, with a small hole drilled vertically through the middle of each step. Once the pin is through the insect, place the pin point in the hole of the 1in step and push it down until it touches the hard surface upon which the block rests. With thick-bodied insects you may need to withdraw the pin somewhat to leave space enough at the top for easy handling.

Mounting small insects

Some insects are so small that pins would damage their delicate structures. Wide-bodied insects less than 6mm long or narrow-bodied specimens less than 9mm in length are best mounted on points, small triangular pieces of thin white cardboard made entirely from rags so that they will not deteriorate with age. Special punches to make points may be purchased from biological supply houses, as can the points themselves, or you can make your own by cutting out isosceles triangles 4mm at the base and 8mm high. Bend the tip of the point slightly downward with forceps and use a very small amount of white glue or clear fingernail polish to attach the right underside of the insect to the tip of the point, trying not to obscure many of its features. The point can then be positioned on a pin, using the 1in step of the pinning block.

Spreading board

Butterflies and moths must be pinned with their wings fully spread, as these usually display the most important identifying characteristics. Dragonflies and damselflies are mounted in a similar fashion, and grasshoppers may have one of their wings spread to show their detail. A spreading board has a groove or slot in the center to accommodate the body of the insect, and two

slightly sloping sides on which the wings are positioned with pins and strips of paper as they dry. This gives the wings a slight upward tilt, allowing them to settle approximately to a flat plane as they age. Specimens should be pinned before their wings are spread, and you will find a cork strip at the bottom of the central groove to accept the pin on which the insect is mounted. Wings of butterflies and moths should be arranged so that the rear edge of the forewing is perpendicular to the body. Drying times depend upon the humidity, but generally vary from two weeks for smaller insects to one month for larger ones.

Alcohol preservation

Some insects are simply too small to mount even on points, and others have soft bodies that would wilt if mounted by conventional means. These are best preserved in small, tightly-capped vials containing a 70 per cent solution of alcohol. Several members of the same family, collected at the same time and place, may share a vial. Replace the alcohol once about a week after introducing the insects. Vials with leak-proof caps can be secured horizontally in display cases with the rest of your collection, but they may need to be recharged with alcohol occasionally, as small amounts may evaporate over time.

Labeling specimens

Unlabeled specimens have very little value to the scientific community. At the very least, each specimen should be labeled with the location and date of its capture and the collector's name. Location details should include the country (abbreviated), state (abbreviated), and county or nearest town. Record the date with the month abbreviated (for example, Mar. 4, 1991) rather than written numerically (3/4/91), which could be mistaken for April 3, 1991. If your full name is too long to fit on the label, it may be recorded by your first two initials plus last name. Use the 18mm step on your pinning block to position this label parallel to the insect's body (head-to-toe axis) with the pin in the center of the label.

Additional information that would add to the value of your specimen may be recorded on a second label. Such data may include the collector's name (if this is not on the first label), and the insect's habitat, the method of capture or equipment used, the elevation at which it was found, or

Above A particularly fine example of a glass-fronted insect cabinet; now a collector's piece, this is the best possible way of preserving a collection of insects. The moths seen here are displayed in species rows, each specimen carrying a label of name, date, and locality. Every so often the drawers must be filled with an insecticide to deter the museum beetle, which can decimate collections in a few months.

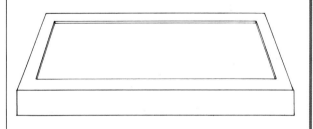

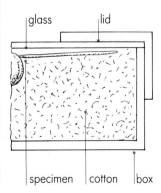

glass | lid

specimen | cotton | box

Above and **left** Pinned insects should be kept in a box or case with an airtight lid. There are a number of models available including this Riker mount. Here, pinned specimens with spread wings are placed upside down on cotton just below the glass.

Right The French naturalist, Jean-Henri Fabré – frequently called "the insect man" – collected plants and animals, but especially insects. He was a contemporary of Charles Darwin and lived in Provence, France. Today his collection can be seen just as he left it. His study has numerous boxes filled with local insects from the South of France and from the Massif Central.

a specific landmark. This label should be placed parallel to the first label and underneath it, using the 12mm step on your pinning block.

Each of these labels should measure 8.5mm × 17mm This is a fairly small area, so you must print very small, preferably with an extra-fine point pen and permanent ink, such as India ink. Use stiff white paper, with 100% rag content, that will not deteriorate.

If you are certain of the specimen's identity, you may record the species name, the name of the person who identified it (preceded by "det." for "determination"), and the year in which it was positively identified. This data should appear on a label measuring 25mm square and located on the pin on the floor of the display case.

Caring for your collection

A collection in which you've invested valuable time deserves a good home. This should be a flat case, preferably wooden, with a snug lid that is as airtight as possible and closes securely. Various models are available through biological supply companies, and as your collection grows, you might want to consider buying a cabinet into which display cases fit as drawers. If you are fortunate enough to know a talented woodworker, ask that person to construct a display case for you, preferably with a large glass window in the lid. Weatherstripping can be used to keep in fumigants

and keep out dust and insect pests that feed on specimens. Line the bottom with soft, white cardboard over cork or polystyrene to accept pins.

Collections should be arranged by orders, and by families within each order. Orders should be arranged from the primitive to the most modern, as determined from your field guides. You can purchase or make individual unit trays from white cardboard to accommodate separate orders, or even families in larger collections. A large label with the Latin and common names of each order may be pinned to the floor of the case, with a label listing the Latin and common names of each family pinned below the order label. You may wish to color code the ink or paper on order and family labels for display purposes. Arrange specimens of each family in orderly rows and columns.

Pests such as dermestid beetles that feed upon dry specimens can be repelled by including fumigants, such as naphthalene or para-dichlorobenzene (PDB), both sold commercially as either moth balls or moth flakes. A combination of the two works well, as PDB is more toxic to pests but naphthalene lasts longer. Place them in small boxes with screened lids in the corner of your display case, and check them every month or two. If an infestation does occur, as evidenced by sawdustlike material beneath specimens, or shed beetle larva cases, introduce ethyl acetate on a wad of cotton or gauze to the closed display case.

INSECT PHOTOGRAPHY

A challenging and rewarding alternative to keeping an insect collection is insect photography. For practical entomologists, there is only one group of cameras to seriously consider; the 35mm single-lens reflex (SLR) models *with interchangeable lenses* offer by far the best compromise between quality, versatility, convenience, ease of operation, and cost. These are the first choice of nature photographers around the world.

The quality of photographs taken by a competent photographer using a 35mm SLR camera is difficult to beat, and its versatility is second to none. With it, possible subjects range from the microscopic to the astronomical. By simply changing lenses, you can go from shooting close-ups of a beetle to taking telephoto shots of a skittish songbird in a matter of seconds. They are also light enough to go anywhere. Their ease of operation varies, but to use the newest auto-focus, auto-exposure cameras is literally as simple as point-and-shoot. Finally, the cost of 35mm cameras and basic accessories is well within the budget of most people. Second-hand cameras in good condition can often be found in camera shops or through classified advertisements, but it is wise to have any second-hand equipment examined by a qualified camera technician before purchasing it.

Choosing a brand of equipment is largely a matter of personal preference. For background information, write to the major photography magazines and request their field test reports on any models you are considering. These magazines also publish annual buyer's guides, which are helpful in revealing what is available. You should pay attention to the quality and variety of lenses and accessories available for different brands, but there is no need to buy top-of-the-line equipment to get good results. Remember the story of the photographer and the writer who met at a party. "I saw your exhibit at the gallery last week," said the writer. "You must have a very good camera!" To which the photographer replied, "Why, thank you. By the way, I really enjoyed your latest book. You must have an excellent word processor!"

Certain accessories are quite useful to the practical entomologist, while others have limited application in this field. Lenses are labeled by their focal length, which in turn determines the

Features to look for when selecting a camera

Mandatory
Single-lens reflex
Interchangeable lenses
Depth-of-field preview
Built-in light meter
Self-timer

Optional
Auto-exposure with manual override

Auto-focus with manual override
Motor drive
Dedicated TTL flash capability
Auto-rewind
Lock-up mirror
Interchangeable focusing screens
Built-in flash

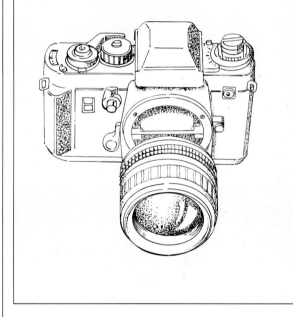

perspective they yield. Most cameras are offered for sale with a standard 50mm lens, although you may wish to purchase the camera body alone and select your own lenses. A 50mm lens gives roughly the same perspective, or angle of view, as that of the human eye.

Lenses

Wide-angle lenses are those with a focal length of less than 50mm. These provide a wider field of view than seen by the human eye, and consequently the subjects appear smaller, although the difference is negligible until you get down to a focal length of 35mm or less. Wide-angle lenses are useful for taking photos of landscapes and habitats, and also for photographing large subjects

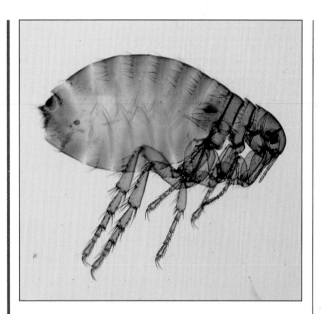

at a close distance. They have a limited application in photographing insects.

At the other extreme are telephoto lenses, those with focal lengths ranging from 50mm to 1,000mm or more. However, only the medium-length telephoto lenses, up to approximately 200mm, have any practical application in photographing insects, and then only when used with close-up accessories that will be discussed later. Telephoto lenses provide a narrower field of view and subjects appear larger than when seen by the unaided eye. They are used primarily when you cannot get as close to your subject as you wish, or in order to isolate the subject and remove a cluttered foreground.

High-quality zoom lenses, providing variable focal lengths, are extremely practical for field work. The entire range of focal lengths from about 24mm up to 210mm can be covered with just two lenses. This leaves you relatively unencumbered and able to carry other gear as well. Adjusting the focal length allows you to compose a photograph exactly as you want it without having to move the camera closer to or farther from the subject.

Macro accessories

Unlike most other fields of nature photography, close-up equipment is necessary for insect photography. Macro lenses can be used at their designated focal length, but also allow you to continuously focus down to very close distances, usually a matter of a few inches. They can thus do double duty, and many photographers prefer to purchase a 55mm macro lens instead of the

Top left Results like this can be achieved with some experimentation and a special camera attachment for a microscope. The insect can then be magnified many times, and photographed with a 35mm SLR.

Left This close-up of a butterfly proboscis was taken with a 50mm lens, three extension tubes and a tripod. Close-up equipment enables you to photograph in minute and fascinating detail, but it may take some practice before you become expert.

standard 50mm lens offered with their camera body. Macro lenses are available in a variety of focal lengths, and the shorter the focal length, the closer you must be to the subject in order to achieve maximum magnification. The best macro lenses give 1:1 reproduction, which means that the image on the frame of film is the same size as the actual subject.

Results equal or superior to those possible with macro lenses can be achieved with standard lenses by adding space between the camera body and the lens, usually by means of extension tubes or bellows. Extension tubes are less expensive and more practical to use, but offer somewhat less versatility in achieving a specific image size.

With the aid of an adapter, you can convert your ordinary lenses into high-quality macro versions by reversing them on the camera body. Similarly, you can purchase adapters to reverse one lens on top of another to achieve super-magnification. One drawback to both methods is that the aperture on the reversed lens must be set manually, negating any auto-exposure capability your camera may have. Also, by linking two lenses you will dramatically reduce the amount of light reaching the film, which means that you will almost certainly need to use a flash.

One close-up accessory that is a great convenience, if not a necessity, is a focusing rail. The range of distances from the camera that is in crisp focus, called *depth-of-field*, is reduced proportionately to the magnification being used. This means that when using the high magnification possible with some macro equipment, it is often difficult or impossible to hold a camera steady enough to keep what you want in focus. By mounting the camera on a focusing rail and securing that on a tripod, you can make fine adjustments in camera position.

Tripod

One of the most effective accessories you can use to improve the quality of your photographs is a sturdy tripod. Most blurred images result not from poor focusing but from camera movement. The steadiest person in the world cannot hold a camera as still as a mediocre tripod. Although a tripod can improve sharpness at any shutter speed, the universal rule of thumb is not to hold a camera when shooting at less than the reciprocal of the focal length of the lens. Put more simply, when

Top When using a large aperture (f2.8) it is important to focus carefully on your subject as the foreground will be slightly out of focus.

Above When using a small aperture (f.22) the depth of field increases and the overall focus is sharp.

Guidelines for purchasing a tripod

• Buy the best tripod you can afford, best in this case meaning sturdiest, not the fanciest. Before you buy, set up the tripod and lock everything in place, then wiggle, twist, and turn every part of it. Nothing should move – not at all.

• Select a tripod that allows you to shoot at or near ground level, but avoid those with reversible center columns. These do indeed permit you to shoot at ground level, but oblige you to work with a tripod leg between you and an upside-down camera, a nearly impossible task for most of us.

• Buy a tripod that reaches your eye level *without* the center column extended, which would effectively convert your tripod into a less stable three-legged monopod.

• Make sure the tripod and camera are light enough for you to carry them without undue strain. The strongest tripod in the world is completely useless if it has to be left at home because it is too heavy.

• Buy and use a cable release with your tripod. This will prevent camera movement caused by depressing the shutter button with your finger.

Above Here, a camera is mounted on a tripod, and the 35mm lens is separated from the camera body by bellows (a close-up accessory); the cable release trips the shutter without blurring the image.

using a 50mm lens, you should use a tripod or other steady support if your shutter speed is less than 1/50 of a second. Hand-holding a 200mm lens requires a shutter speed of at least 1/200 of a second, and so on.

When you need a tripod but can't use one because your subject doesn't stay in one place very long, a shoulder stock may solve the problem. This permits you to steady the camera against your shoulder and aim it like a rifle. The grip even has a trigger connected to a cable release that trips the shutter. A monopod is also useful for impromptu occasions.

Films

Available films vary nearly as much as camera models. What is right for you depends upon how you want to use your images, and the conditions under which you are shooting. Print films are developed into negatives, from which a print is made on light-sensitive photographic paper. Prints are easy to display and view, whether in an album, or framed on a wall, or otherwise. Black-and-white prints will produce an accurate representation of the form and texture of your subject, but they cannot portray the vast array of colors among insects. Color prints are more aesthetically pleasing and provide a more accurate representation of natural subjects than do black-and-white prints, but they are also the most expensive types of film to develop, thus limiting those on a tight budget.

Color slide films, also known as color reversal films, are the workhorse of professional nature photographers, especially those shooting for publication, since transparencies (and black-and-white prints, to a lesser extent) are used almost exclusively in the publishing business. Transparencies cannot be viewed easily without a projector, slide viewer, or a loupe and light table. Their major application, aside from publishing, is their use in illustrating lectures. Slide film is much less expensive to develop than color print film, and you can also have prints made from slides at a custom photo laboratory. Although it costs more per picture than print film, you can afford to experiment with different exposures and compositions and print only those you like.

The other major variable associated with films is light sensitivity, better known as film speed and denoted by an ASA (American Standards Association) number. The higher the ASA number, the

faster, or more light sensitive, is the film. Slower films therefore require a slower shutter speed *or* a larger lens opening under a given light intensity than do faster films.

Greater light sensitivity is achieved by making the grains on the film emulsion larger. Photographs produced with faster films are consequently more grainy than those produced with slower films. In other words, the lower the ASA number, the sharper the image. For this reason, publishers generally will not consider using images produced on film rated higher than 100 ASA. High speed films are most useful under low light conditions, if you have chosen not to use a flash, or when you want to use a very fast shutter speed to freeze movement.

Exposure

A photograph records light. In order to achieve the correct exposure, the correct amount of light must reach the film. This is a function of the combined shutter speed and size of the lens opening, called the *aperture* or *f-stop*. Modern 35mm cameras have built-in light meters which measure the amount of light reaching the film. Most cameras can use this information to calculate the correct aperture for a given shutter speed, or vice versa, and some will calculate both.

Making a correct exposure is very much like trying to fill a bucket exactly to the top with water from a faucet. A slow trickle takes longer to fill the bucket than opening the faucet all the way. If too much water is added, the bucket overflows; with too little water, it is not as full as you wish. Correspondingly, the correct exposure of a photograph requires the use of a longer shutter speed with a small aperture than with a large one, and vice versa. Too much light results in overexposure with washed-out highlights, and too little results in an underexposed image.

Natural light vs. electronic flash

Whether or not you use an electronic flash will depend on several factors. If you want your photograph to look as natural as possible, try to use whatever light is available at the scene. Bright overhead sunlight will result in harsh, unappealing shadows, but on smaller subjects you can fill these in by using reflectors made by covering cardboard with crinkled aluminum foil. Shadows may also be softened by constructing a light-diffusing lean-to

Top Because the bees are not moving quickly a shutter speed of 1/4 second has proved suitable. At such slow speed a tripod is essential.

Above Photographs taken in strong sunlight often result in harsh shadows. Using a flash has softened the contrasts while highlighting the shiny bodies of the bees.

from translucent sheets of plastic, such as those used to cover fluorescent light fixtures.

Low natural light will require that you use a large aperture, a slow shutter speed, or both. Slower shutter speeds demand the use of a tripod to attain sharp focus, but a constantly moving subject will effectively prohibit the use of slow shutter speeds, even with a tripod. In cases like this, or when you wish to darken the background and emphasize the subject, electronic flash is the best choice. Using flash also allows you to fill in harsh shadows, freeze movement, or use a smaller aperture to

Top Fast films are useful in low light conditions where it is undesirable to use a flash. However, the image appears grainy, as in this 400ASA picture.

Above Slower films require more lighting but produce a sharper image, as in this 100ASA picture.

need not be powerful since it is so close to the subject.

Another accessory, the ring flash, eliminates the need for a specialized bracket. A ring flash is circular and attaches to the front of your lens. While it produces consistent results, the even illumination that it produces tends to wash out shadows and may produce photos that lack a sense of depth. Also, any highlights reflected within the photograph will appear as small circles rather than natural-looking points of light.

Focus

Another function of aperture size is depth-of-field, the area within a certain range of distances from the camera which is in focus. Using a given lens, a smaller aperture will yield a greater depth-of-field than a larger aperture. Should you desire to emphasize your subject by focusing on it rather than on the background or foreground, you would select a larger aperture and a faster shutter speed. If you desire more depth-of-field, to portray a plant in its natural habitat for instance, you select a small aperture, but to get the correct exposure you must also employ a correspondingly slow shutter speed. The focal length of a lens also affects the depth-of-field. Shorter focal lengths yield greater depth-of-field than longer focal lengths. Most lenses have a scale that will show your depth-of-field in feet and meters from the camera, so you can calculate which parts will be in focus.

increase depth-of-field, a concept we shall discuss shortly.

The most aesthetic electronic illumination of a small subject is produced when the flash is positioned close to that subject and above the camera lens, producing soft shadows and dupli-cating the natural angle of illumination from the sun. To achieve consistent results when using lenses of different focal lengths, you will need to buy or make a bracket that allows you to position the flash over the same reference point, such as the end of the lens, every time. The flash itself

Recommended photographic equipment	
Option 1	**Option 2**
35mm SLR camera body	35mm SLR camera body
55mm macro lens	28–70mm zoom lens
105mm macro lens	70–210mm zoom lens
200mm macro lens	Extension tubes/bellows
Extension tubes/bellows	Focusing rail
Focusing rail	Warming filters
Warming filters	Polarizing filters
Polarizing filters	Electronic flash
Electronic flash	Tripod
Tripod	Cable release
Cable release	Reflector
Reflector	Diffuser
Diffuser	

KEEPING A FIELD NOTEBOOK

Insect field guides and identification keys are wonderful tools that provide excellent background information. However, *you will learn and remember more about insects from your own detailed observations than you will from reading about them!* For this reason, a meticulously maintained field notebook is by far your most valuable learning tool. Notes about anything that strikes you as interesting or significant will serve you for the rest of your life.

A field notebook is not a diary. You need not make an entry every day, week, or at any regular intervals. To get the most from a field notebook, however, it should be part of the basic gear that you take along every time you go into the field, whether that be for collecting, photographing, or just taking a hike. If your interests include other subjects besides entomology, these also have a place in your field notebook. Basic information in each entry should include the date, departure and return times, weather conditions, temperature, and geographic locations and habitats visited.

Always record your observations on the spot; important details will be forgotten if you trust them to memory. Even a short note with a few key words will help jog your memory and allow you to fill in the details later. Also, regardless of your artistic abilities, always include many sketches with your notes. No matter how crude, sketches are great memory triggers that help you to visualize what you saw while making the sketch. Sketching also trains you to make precise observations, a skill critical to any naturalist. You may find that drawing only a portion of the organism helps to focus your observations and reveals greater detail.

Logbook
As a supplement to your field notebook, consider keeping a loose-leaf logbook at home, into which you can transcribe your notes, reorganize them, expand them, and co-relate them with your readings and past observations. Do this without fail after each excursion, and before you know it you will have amassed an impressive body of knowledge, some of which may not be found in field guides and, in fact, may never have been recorded.

Catalogue card file
In addition to, or in place of, a logbook, you may find it useful to start an index card file on the families and species that you have come to know. Use one card per discovery and record its name, date found, relevant information, and the page number of your field notebook on which you make reference to it. Number each card and use the corresponding number to identify photographs, specimens, and notebook entries. It is easier to have your notes on file in this manner than to search through notebooks for information.

Don't expect to learn everything about a particular species the first time you find it. Think of each encounter as a jigsaw puzzle with the pieces all askew. With each encounter, you have the opportunity to put a few more pieces into place. In time, an image begins to emerge, and the background of the picture also becomes more distinct as you learn more about how the organism interacts with its surroundings. Keeping a field notebook and a logbook or card file enables you to save and assemble all the pieces of the puzzle.

Typical Catalogue Card

Notebook page number

Specimen number
Common name
Scientific name
Order
Date
Location
Habitat
Description

Family

KEY TO INSECT ORDERS

The following key will give you a starting point from which to identify any unfamiliar insect by placing it in the correct order. Begin with number one and decide which of the two options, a or b, fits the insect in question. Whichever one you choose will either reveal the order or direct you to another part of the key, and so on, until the order is disclosed. For best results, it is important to read each question carefully and examine the insect meticulously. Some orders have members that are both winged and wingless, and hence are listed twice in the key.

1 Does the insect have wings?
a. Yes .... Go to #2
b. No .... Go to #17

2 How many pairs of wings does the insect have?
a. One .... Diptera
b. Two .... Go to #3

3 Do the two pairs of wings differ greatly in structure, the first pair being thick and hard or leathery?
a. Yes .... Go to #4
b. No .... Go to #7

4 Is the first pair of wings rigid, and do they meet in a straight line down the middle of the back?
a. Yes .... Go to #5
b. No .... Go to #6

5 Is there a pair of prominent pincerlike cerci at the tip of the abdomen?
a. Yes .... Dermaptera
b. No .... Coleoptera

6 Does the insect have:
a. chewing mouthparts, front wings leathery and heavily veined, and hind wings folded like a fan? .... Orthoptera
b. sucking mouthparts and front wings leathery at the base, membranous and overlapping at the tip? .... Hemiptera

7 Are the mouthparts a coiled tube and the wings covered with scales?
a. Yes .... Lepidoptera
b. No .... Go to #8

8 Are the wings rooflike, sloping downward and outward from the middle of the back?
a. Yes .... Homoptera
b. No .... Go to #9

9 Is the insect slender and mothlike, with long, slender antennae and wings that are widest past the middle?
a. Yes .... Trichoptera
b. No .... Go to #10

10 Do the wings have few or no cross veins?
a. Yes .... Go to #11
b. No .... Go to #12

11 Does the insect have chewing mouthparts and hind wings somewhat smaller than the front wings?
a. Yes .... Hymenoptera
b. No .... Thysanoptera

12 Are there two or three long, slender, tail-like appendages on the tip of the abdomen?
a. Yes .... Ephemeroptera
b. No .... Go to #13

13 Does the head have an elongated trunklike beak with chewing mouthparts at its tip?
a. Yes .... Mecoptera
b. No .... Go to #14

14 Does the insect have inconspicuous antennae, long narrow wings, and a long slender abdomen?
a. Yes .... Odonata
b. No .... Go to #15

15 Does the insect have two short cerci on the tip of its abdomen and front wings narrower than the rear wings?
a. Yes .... Plecoptera
b. No .... Go to #16

16 Do the tarsi each have 5 segments?
a. Yes .... Neuroptera
b. No .... Isoptera

17 Is the insect antlike, with a narrow waist?
a. Yes .... Hymenoptera
b. No .... Go to #18

18 Is the insect antlike, but with a wide waist?
a. Yes .... Isoptera
b. No .... Go to #19

19 Is the insect small and flattened, with chewing mouthparts and a head about as wide as its body?
a. Yes .... Go to #20
b. No .... Go to #21

20 Are the antennae long, and composed of many segments?
a. Yes .... Psocoptera
b. No .... Mallophaga

21 Is the insect's body soft and rounded, with two short tubes protruding from the abdomen, and with a small head?
a. Yes .... Homoptera
b. No .... Go to #22

22 Is the insect very small, with a vertically flattened body, a hooklike claw on each leg, and sucking mouthparts?
a. Yes .... Anoplura
b. No .... Go to #23

23 Is the insect very small and narrow (flattened laterally) with sucking mouthparts?
a. Yes .... Siphonaptera
b. No .... Go to #24

24 Is the insect:
a. delicate with chewing mouthparts and threadlike "tails" and antennae? Thysanura
b. very small with a springlike lever folded under its abdomen which it uses for leaping? .... Collembola

DRAGONFLIES & DAMSELFLIES
ORDER ODONATA

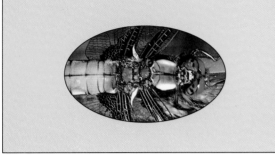

INTRODUCTION

Dragonflies and damselflies, large and often brightly colored, are arguably the most accomplished aerialists in the animal kingdom. Dragonflies, in fact, are not only remarkably swift and agile, able to travel at up to 35 miles per hour and reverse direction in midair within one body length, but they can also hover with ease and fly backward like tiny helicopters. Their wings are reinforced with deep corrugations at the base, which accounts in part for their strong flight.

Most of the stronger insect fliers have evolved means to link their fore wings and hind wings in flight so that each side functions as one flight surface. Dragonflies and damselflies, in contrast, move the two pairs independently, timing the stroke of the hind pair so that they meet the oncoming air before it has been disturbed by the front pair; physicists would describe them as operating in *antiphase*. Even more remarkable, dragonflies have evolved a technique that endows them with tremendous power for their size. Researchers studying dragonflies in a wind tunnel have found that dragonflies twist their wings on the downstroke, creating miniature whirlwinds that move the air much faster over the upper wing surface, reducing the air pressure there and greatly increasing lift. These findings are deemed to have so much potential value in aviation that

the United States Navy and Air Force have contributed to the funding of such research.

Damselflies and dragonflies catch their prey on the wing. Their long, slender legs, each with a row of stiff bristles on either side, are held slanting forward in flight to form a basket that scoops smaller flying insects out of the air and traps them until the jaws can grasp them. Their sharp chewing mouthparts enable them to cut up their prey into bite-sized pieces. They are quite beneficial to humans in that they consume huge numbers of mosquitoes and black flies.

The success of these "dragons of the air" as aerial hunters is due largely to their acute vision. Their huge compound eyes, each composed of 10,000–30,000 individual facets, are supremely adapted for detecting motion, and their heads have an unusually large range of motion for insects, enabling them to be cocked at different angles and perhaps improving their depth perception from different perspectives.

Dragonflies and damselflies copulate in the air, and you can observe these mating flights over any body of fresh water in mid-summer. Each male curls the end of his abdomen forward to place a packet of sperm in a cavity underneath his second abdominal segment, then grasps a mate by her neck with clasping cerci at the tip of his

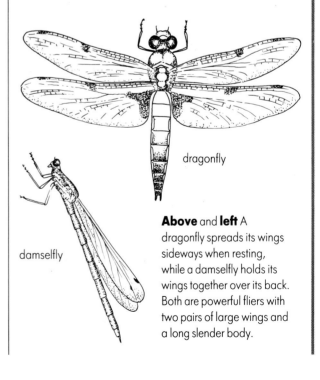

dragonfly

damselfly

Above and **left** A dragonfly spreads its wings sideways when resting, while a damselfly holds its wings together over its back. Both are powerful fliers with two pairs of large wings and a long slender body.

What you can do

Dragonfly watching

Though it will never attain the popularity of bird watching, observing dragonflies and damselflies has its own rewards. By employing slow, steady movements on land or using a drifting canoe or rowboat, one can approach these wary insects quite closely and observe territoriality, mating, egg-laying, predation, and even metamorphosis, all within a relatively small area.

Above Freshly-emerged, an adult darner dragonfly (*Aeshna juncea*) dries its expanded wings. Its body and wings are still a downy light color that darkens as the insect dries.

abdomen. In flight, she curls the tip of her abdomen forward to retrieve the sperm capsule. She later deposits her fertilized eggs in or near the water.

Naiads

Odonatans undergo hemimetabolous development, but with an interesting twist. The immature insects, called *naiads*, are aquatic and bear little resemblance to the adults. They breathe by means of tracheal gills lining the rectum, and damselfly naiads also have three long tail-like gill plates extending behind the abdomen.

Naiads of this order are among the chief aquatic insect predators. They lurk stealthily among vegetation or lie quietly concealed by silt and debris, but dragonfly naiads can dart forth with amazing speed, thanks to their jet propulsion system that shoots a powerful stream of water from the tip of the abdomen. Another weapon at their disposal is an elongated hinged lower lip, or labium, with a pair of grasping jaws at the tip. While usually folded under the head with the scoop-shaped tip masking the face, it can shoot forward with lightning speed to snatch prey an impressive distance ahead.

When they are ready to become adults, damselfly and dragonfly naiads crawl up plant stems, sticks, dock pilings, or any other structure protruding from the water. Here, each undergoes its last molt and flies away, leaving the cast skin, or *exuvium*, still clinging to its perch, a common summer sight around lakes and ponds.

Along with mayflies, dragonflies and damselflies are the most ancient of flying insects. Their ancestors, some of which developed wingspans of up to 30 inches first appeared during the Carboniferous Period, between 350 and 280 million years ago. Today there are about 5,500 species of dragonflies and damselflies worldwide, with approximately 450 species in 11 North American families. They are usually found over water, where males will defend territories against rival males of the same species. Members of this order are easily recognized by their bulging compound eyes, heavily-veined wings, and long, slender abdomens. They cannot fold their wings flat against their back, as can many other orders. Instead, dragonflies rest with their wings to the sides, perpendicular to the thorax, and damselflies hold their wings vertically and rearward.

69

DRAGONFLIES

Darners – Family Aeshnidae

Darners are the largest and swiftest of North American dragonflies. Adults range in size from 50–120mm in length and may have wingspans up to 150mm. Brilliant green, blue, or brown, with clear wings, they are easy to recognize, being the only family of dragonflies whose compound eyes meet along the top of the head. Darners are also the dragonflies most likely to be found fair distances from water, although they usually haunt ponds and slow streams. They are named for their resemblance in flight to darning needles.

Clubtails – Family Gomphidae

Named for their swollen, terminal abdominal segments, clubtails favor streams over ponds. They soar and hover much less than other dragonfly families, preferring instead to rest on logs, stones, and leaves. They catch prey while darting from one resting place to another. The bodies of adult clubtails are 31–90mm long and darkly-colored with yellow or green stripes. Identifying features include their widely separated eyes and spreading legs.

Biddies – Family Cordulegastridae

A small family, biddies haunt small woodland streams almost exclusively, cruising slowly about 12 inches over the surface. Adults are 45–88mm long and smoky brown, with yellowish markings and a hairy appearance. Their compound eyes meet or are slightly separated at a single point only.

Green-eyed Skimmers – Family Corduliidae

Adults of this family are 28–78mm long and usually black or metallic, with bright green eyes, although some are dark brown with yellow markings. The hind margin of each compound eye is distinctly lobed. Green-eyed skimmers favor woodland swamps and ponds.

Belted Skimmers and River Skimmers – Family Macromiidae

Belted skimmers are brown with lighter markings. They are uncommon, but usually occur along the marshy margins of ponds. River skimmers, denizens of rivers and slow streams, are blackish

Above The hairy thorax of this golden-ringed dragonfly (*Cordulegaster boltoni*) is obvious, indicating that it is freshly-emerged from the naiad. The naiads breed in muddy moorland streams or in boggy areas.

Above The four-spotted chaser dragonfly (*Libellula quadrimaculata*) inhabits boggy pools and ponds. Males may be territorial, defending their own patches.

Left The wide and comparatively short body of the broad-bodied chaser dragonfly (*Libellula depressa*) is typical of this group of dragonflies, and in this female the orange markings are highlighted on the base of the wing veins. The articulation joints are clearly visible on the wing bases.

Above The tail of this female club-tailed dragonfly (*Gomphus vulgatissimus*) is a key identification feature. A member of the Gomphidae family, it has a characteristic small, heart-shaped head. The naiads breed in slow-flowing streams and rivers.

with yellow markings. Both range from 56–91mm in length.

Common Skimmers – Family Libellulidae

Common skimmers are the largest family in this order and are represented by the most common dragonfly species. They have brightly colored bodies, 18–75mm long, that are noticeably shorter than the wingspan. The males and females of some species are colored differently. Most dragonflies whose wings have colored bands or spots belong to the Libellulidae. They're fond of perching on plants along the margins of ponds, lakes, marshes, and quiet streams. The flight of common skimmers is swift, sometimes interrupted by intervals of hovering.

What you can do

Underwater jets

To observe the unique propulsion system of dragonfly naiads, set up an aquarium with about 8cm of muddy water and allow a thin layer of pond silt to settle on the bottom. Introduce one or more immature dragonflies; prod them gently with a blunt stick to get them moving, and you will see a stream of water shoot from the tip of the abdomen and stir up silt as the dragonfly darts forward. This is primarily an escape mechanism, as the naiad relies upon stealth to capture prey.

DAMSELFLIES

Broad-winged Damselflies – Family Calopterygidae

A striking family, the broad-winged damselflies often display bright metallic bodies, 25–51mm in length, and wings that are dark or have brightly colored spots. Though their wings narrow at the base, they are not stalked, as are those of other damselfly families, hence their name. They tend to frequent woodland streams, often perching on overhanging vegetation.

Narrow-winged Damselflies – Family Coenagrionidae

Narrow-winged damselflies include more species than any other damselfly family. These brightly colored insects, measuring 25–50mm in length, prefer quiet waters, such as ponds, marshes, and swamps. Colors and patterns usually differ bet-ween the sexes, with males being the more colorful of the two. Most have clear wings, extremely narrow or stalked at the base, that are held together vertically over the body when at rest

Spread-winged Damselflies – Family Lestidae

The clear wings of this family are also stalked, but unlike those of the narrow-winged damselflies, they are held partly spread over the body at rest, instead of vertically together. Ranging in length from 32mm to 62mm, members of this family are most often found around the margins of pond, swamps, and marshes.

Below The banded demoiselle (*Calopteryx splendens*) stands delicately on a leaf with its tarsi and claws while showing off its bristly legs. This male is ready to attack.

Right This side view of a large red damselfly (*Pyrrhosoma nymphula*) reveals its large thorax, which houses the powerful flight muscles, as well as the long abdomen.

What you can do

Tips for collecting

Capture techniques
aerial net

Killing method
ethyl acetate killing jar

Preservation techniques
pin on spreading board

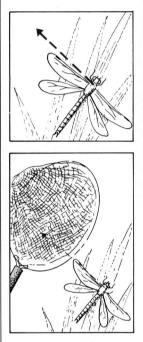

2 Place dragonflies and damselflies in triangular paper envelopes for protection, with their wings folded above the body.

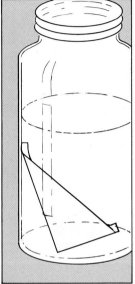

1 Dragonflies are so adept at evading other aerial predators that obtaining specimens can prove quite frustrating, even with a net. The secret is to remember that resting dragonflies take off at a 45 degree angle forward when alarmed, so wait for your specimen to land; approach slowly; place the net in the projected path of escape, and startle the dragonfly.

3 The bright colors on many specimens tend to fade, but this can be minimized by rapidly drying them in the sun, an oven, or under a lamp, after they have been pinned. You can also preserve their colors by keeping them in the paper triangles and sealing these in a jar of acetone for 24 hours, then allowing them to air dry. Acetone is flammable, so use caution with this method.

GRASSHOPPERS, KATYDIDS, & CRICKETS
ORDER ORTHOPTERA

INTRODUCTION

The Order Orthoptera includes most of the champion jumpers of the insect world. Proportionately, were we as good, we could clear over 300 feet in the high jump and 500 feet in the broad jump! Jumping high, far, and fast is their primary defense against predators, so those that leap best also survive longer and reproduce more successfully, creating even better jumpers in future generations. Characteristically, most families have greatly swollen femurs on their hind legs, housing powerful jumping muscles. Many species in this order also rely heavily on camouflage for protection.

There are about 1,200 North American species of these medium-to-large insects in 10 families, and about 20,000 species worldwide. Orthopterans usually have narrow, leathery forewings and broad, membranous hind wings that fold accordion-style under the forewings when at rest.

Orthopterans undergo hemimetabolous development. Both the nymphs and the adults have chewing mouthparts and are mainly vegetarians, although some cannibalism has been observed when high populations deplete the available food supply. Some are omnivorous, and others, especially those that live underground or within human dwellings, are scavengers, eating decaying plant and animal matter.

Sound is very important in the courtship of most orthopterans. Consequently, they have some of the most well-developed tympanic organs among insects. These sensitive structures are located either on the tibia of each front leg or on each side of the first abdominal segment.

Depending upon the family, their songs, called *stridulations*, are produced either by rubbing their spiny hind femurs against a projecting vein on the outside of the forewing (*tegmen*) or by rubbing a hard scraper on one fore wing against a file-like vein on the other, much like one would draw a bow over violin strings. Each song is unique to an individual species, so there is no mistake as to who is calling. It is the males that sing, and the females, recognizing the song of their species, seek out the serenader.

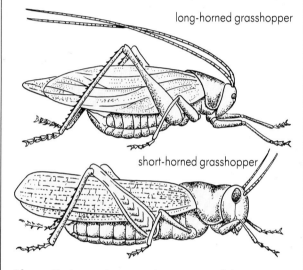

long-horned grasshopper

short-horned grasshopper

Above Crickets and grasshoppers are noted for jumping and singing. Most families have swollen femurs housing powerful jumping muscles, and some of the best developed tympanic organs in the insect world.

What you can do

Collecting tips for the Order Orthoptera

Capture techniques	**Preservation techniques**
aerial net	pin (Specimens with a flash pattern on the hind wing may be mounted with one wing spread.)
sweep net	
hand collecting	
beating tray	
Killing method	
ethyl acetate killing jar	

Left The well-camouflaged Egyptian grasshopper (*Anacridium aegyptium*), from North Africa and southern Europe, has a lugubrious appearance. Its compound eyes are banded like the pattern on the elytra of a Colorado potato beetle.

Above Resting on a tropical zinnia, a bush cricket of the family Tettigonidae (*Tylopsis lilifolia*) has enormous antennae that make it aware of its environment, particularly of scents and touch stimuli.

GRASSHOPPERS & KATYDIDS

Pygmy Grasshoppers and Grouse Locusts – Family Tetrigidae

Tetrigids may occupy many habitats, but they are most often found in damp areas such as stream borders. Though terrestrial, they can swim if necessary. Adults overwinter in sheltered areas and are most often found in spring and early summer. The most notable feature of members of this family is the pronotum, which is long, tapered, and extends backward over most of the abdomen. The forewings are small and may be hidden under the pronotum, but the hind wings are usually well-formed. These are small to medium insects.

Long-horned Grasshoppers – Family Tettigoniidae

Noted for their long, slender antennae, which are as long or longer than the body, long-horned grasshoppers are medium-to-large (14–75mm) insects, and include the well-known katydids. They vary from brown to bright green in color, making them difficult to distinguish from the vegetation they inhabit. Their green colors are absorbed from the chlorophyll in the foliage they eat; in fact, many effectively mimic leaves or other plant parts as a defense against predators. Most prefer the dense leafy cover of trees and shrubs, but some can be found in wet grassy meadows or the borders of streams.

Aside from their antennae, long-horned grasshoppers are also distinguished by the tympanum at the base of each front tibia, by their 4-segmented

Above and **left**: A natural wonder – the mountain grasshopper (*Acripela reticulata*) from Australia. The insect relies on surprise to scare its enemies, showing off its bright red and blue abdomen and the shiny undersides of its forewings in a defensive display designed to deter predators.

tarsi (feet), and also by the swordlike ovipositors used by females to slit plant tissue in order to lay their eggs. Males of this family are among the best songsters of all insects, especially the katydids, whose rasping songs are variations on the tempo of their name, "katy-DID, katy-DIDN'T," and so on. Long-horned grasshoppers overwinter as eggs.

Short-horned Grasshoppers – Family Acrididae

When grasshoppers are mentioned, most people will immediately think of the members of this very common family. These medium-to-large insects (12–80mm long) are named for their short, stout, horn-shaped antennae, which are usually less than half as long as the body. Another notable trait is the location of their large tympanic organs on the sides of the first abdominal segment. Males make a low buzzing song by rubbing the rough surface of the hind wing against the forewing.

Short-horned grasshoppers may readily be found in all terrestrial habitats up to about 14,000 feet in elevation, but are most common in dry grasslands and deserts. They include the numerous species referred to as "locusts." Many attack crop plants. Females exhibit short, stout ovipositors, which they use to lay masses of between 8 and 25 eggs below the ground's surface. Most short-horned grasshoppers overwinter in the egg stage, and do not hatch until summer.

Members of one subfamily of Acrididae, the band-winged grasshoppers, employ flash display, bold colors or patterns on their hind wings that burst into view upon takeoff and startle potential predators. As the insect lands, it quickly stows each boldly-marked hind wing under a dull tegmen and effectively disappears as it drops quickly against a similarly-patterned background. Any bird or mammal in pursuit usually continues to follow the original flight path after the conspicuous wings have folded, overshooting the motionless insect on the ground.

Spur-throated grasshoppers, another subfamily, includes those species most damaging to crops. Some species are migratory and may form swarms that number in the billions.

What you can do

A grasshopper zoo

Equipment

large glass jar with lid
small jar or aluminum can
shoe box
aerial net or sweep net
electric drill (or a hammer and a large nail)

Grasshoppers and katydids are interesting to watch and easy to keep in captivity. They eat plant foliage, and so should be provided with a few leafy plant stems from the area where they were captured, and a bit of damp sponge for humidity.

Turn the lid of the large jar upside down, and glue it to the top of the shoe box lid. After it dries, drill or punch several holes through the lid of the jar, making them large enough to accept medium-sized plant stems but small enough to prevent escape. Next, riddle the shoe box with holes for ventilation. Fill the can or small jar about three-quarters full with water and place it in the shoe box so the ends of the plant stems will rest in it when closed. Place your specimens in the jar; thread 2 or 3 plant stems through the lid; screw the lid onto the large jar; invert the jar and lid on top of the shoe box, and your zoo is finished.

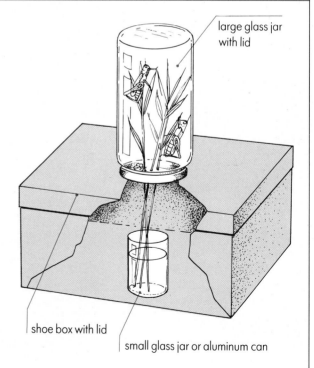

large glass jar with lid

shoe box with lid

small glass jar or aluminum can

CRICKETS

True Crickets – Family Gryllidae

Shorter fore wings, 3-segmented tarsi (feet), and a broad and somewhat flattened body are the principal features distinguishing these medium-sized (9–25mm long) insects from long-horned grasshoppers. They are equipped with long feeler-like cerci on the tip of the abdomen, and females use needlelike ovipositors to lay eggs singly, either in the ground or on plant tissue. Like the long-horns, crickets also have stridulating organs at the base of their fore wings, but the shorter wings of crickets produce a higher-pitched and more musical song. A centuries-old practice in Oriental households is to keep caged crickets in order to enjoy their song.

Cave Crickets and Camel Crickets – Family Gryllacrididae

The members of this family are wingless, nocturnal, and inhabit moist locations under rocks and logs and in caves, trees, and the burrows of other animals. Each of the hind legs includes a stout femur and a long, slender, many-spined tibia. Many common species have a humpbacked appearance, and are gray or tan in color, with a length of between 10mm and 50mm. Few produce sound, the majority having no tympanic organs. Secluded and nocturnal, their sight is not nearly as keen as that of other orthopterans. Instead, they rely upon their very long and supple antennae to feel their way around. They are readily caught in pitfall traps baited with molasses and placed near wooded areas.

Mole Crickets – Family Gryllotalpidae

True to their namesake, mole crickets have scoop-like fore legs, modified for burrowing. Unlike most other orthopterans, their hind legs are not well-developed for jumping. They prefer moist sand and mud, particularly along streams and ponds. Dense coverings of hairlike setae keep damp soil from clinging to their bodies. Not surprisingly, eggs are laid underground, and both the nymphs and larvae feed on plant roots. Despite their subterranean lifestyle, mole crickets can fly well.

Above Because it spends most of its time underground, the mole cricket (*Gryllotalpa* sp.) has a thickened thorax and its first pair of legs are especially thickened for burrowing.

Right A common way of startling predators is through a hind-wing display, such as this performance by a bush-cricket (*Neobarrettia vannifera*).

Left The wood cricket (*Nemobius sylvestris*) is a lively resident of woodlands in south-western Europe, and shares its habitat with the stick insect. Of very nervous disposition, it seeks shelter when disturbed or jumps using its strong back pair of legs. Its long antennae are used to seek for food as well as keeping a general sensory watch on its immediate environment.

What you can do

A cricket thermometer

Equipment
stopwatch, watch with second hand, or digital watch
accurate air thermometer
pencil
field notebook
small flashlight or headlamp

As they are cold-blooded animals, the body temperatures of insects are rigidly governed by the temperature of the surrounding air. Within normal temperature ranges, their rate of activity and metabolism increases and decreases with the air temperature.

Approximate formulas have been worked out to determine the air temperature by the song of the snowy tree cricket (*Oecanthus fultoni*), which is common over most of North America. If T is the air temperature in degrees Fahrenheit and n is the number of snowy tree cricket chirps per minute, then $T = (n-40)/4 + 50$. In other words, to arrive at the temperature in degrees Fahrenheit, count the number of chirps per minute, subtract 40, divide the difference by 4, and add 50. A shortcut to this formula is to count the number of chirps in 14 seconds and add 40.

To arrive at the air temperature in degrees Celsius, use $T = 5/9 (n + 8)$, where n is the number of chirps in 14 seconds. In other words, determine the number of chirps in 14 seconds, add 8, multiply the sum by 5, and divide that product by 9. Who says math is no fun?

The possible variations of this are numerous. Can you correlate the air temperature to the songs of other crickets, grasshoppers, or katydids common in your area? At what temperature do they stop singing? Can the same relationship be demonstrated with other insect behavior (the flashing of fireflies, for instance?)

THE TRUE BUGS
ORDER HEMIPTERA

SELECTED FAMILIES

The word "bug" has evolved into a very general term, variously used to describe any and all insects as well as many other multi-legged creatures, such as spiders, scorpions, ticks, and mites. Technically speaking, though, the only true bugs are the members of the Order Hemiptera.

Hemiptera means "half wing," referring to the characteristic forewings of this order, called *hemelytra*, which are thick and leathery near the base and membranous near the tip, where they overlap. These wings fold flat over the insect's back covering the hind wings, which are the principal flying apparatus and are uniformly membranous and somewhat shorter than the forewings. Between the folded wings is a prominent triangular sclerite called the *scutellum*, the modified dorsal portion of the metathorax.

A third identifying feature of hemipterans is their piercing-sucking mouthparts, housed in a beak-shaped *rostrum*, or labium, that originates far forward on the head and is folded underneath it, pointing backward between the front legs. The rostrum swings out from its resting position as the bug prepares to feed. Precise directional control over this feature has allowed some hemipterans to exploit food sources other than plants, to which the closely related Order Homoptera is restricted.

True bugs may be either carnivorous, preying upon other insects and sometimes small vertebrates, or herbivorous. The mandibles, the outer pair of four thread-like mouthparts housed in the rostrum, have sharp teeth that are used to pierce the outer tissue of the victim. The inner pair of mouthparts, the maxillae, fit snugly together to form a salivary canal and a food canal. Saliva containing digestive enzymes is pumped down through the salivary canal, and the partially digested fluids are sucked up through the food canal. Many of the predators are beneficial to humans, consuming other insects regarded as pests. Predatory bugs frequently have *raptorial* front legs that are short, stout, and adapted for grasping and holding prey as they feed.

For defense, most true bugs have special stink glands that secrete a substance with a disagreeable odor in order to repulse enemies. Among nymphs, these glands are located on the back of the abdomen, while adults utilize glands located on the sides or underside of the thorax. Many advertise this capability with bright colors and bold patterns, recognized by animals that have had an unpleasant encounter with that species before. Others rely on cryptic coloration and employ their stink glands only as a secondary defense. Some true bugs can also deliver a painful bite when handled.

True bugs undergo hemimetabolous development with five developmental stages. Apart from sometimes varying in color and having wings that are reduced or absent, nymphs strongly resemble adults of their species. They develop from eggs that are most often attached to the surfaces of plants or implanted within their tissues. Hemipterans are widely distributed in most habitats, and

true bug

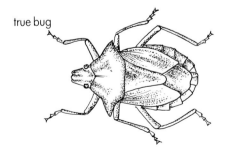

Above Although the term bug is used to describe many kinds of insects, technically speaking only members of the Hemiptera family are known as true bugs. There are over 23,000 known species world wide.

Far left Most aquatic hemiptera possess piercing and sucking mouthparts, but the lesser water boatman (*Corixa punctata*) has reduced mouthparts that are only suitable for eating algae and small animals; it also swims using enlarged and "feathery" legs.

Above left The water scorpion (*Nepa cinerea*) has a "tail" like that of a terrestrial scorpion, which it uses as a siphon for breathing through the water film while it hangs upside down, and lying in wait for its underwater prey.

Left Assassin bugs (reduvids) are true to their name. This one (*Apiomerus falviventris*) from Mexico has caught a bee and is sucking its juices with the same piercing mouthparts that it used to kill its prey.

although most are terrestrial, some very common species are aquatic, either with sprawling legs that distribute their weight over the surface of the water, or with oar-like legs that propel them over or through the water.

Hemipterans are best collected by sweeping and beating vegetation, looking under bark, or using a Berlese funnel or a light trap. Worldwide, there are more than 23,000 species of true bugs in over 60 families, with about 4,500 species in 44 North American families. We will be able to discuss just a few of the more significant families in this chapter.

Water Boatmen – Family Corixidae

Water boatmen are very common aquatic insects and the favorite food of many fish. They are easy to recognize by their elongated oval shape, 3–13mm in length, and by their broad head and eyes, which overlap the anterior end of the prothorax. Their dorsal side is flat, and finely barred with dark lateral lines that produce a mottled camouflage.

Water boatmen swim in a rapid, erratic fashion with oar-like strokes of their middle and hind legs, but most of their time is spent clinging to underwater vegetation, usually in shallow water near the edge of a pond or lake. Periodically, they visit the surface of the water to trap an air bubble, which they wrap closely around their bodies and under their wings, like a silvery envelope, in order to breathe. Their beaks (reduced mouth parts) are short and broad, and useless for piercing and sucking. Instead, they feed on algae, larvae, and other food particles scraped from underwater

surfaces or strained from the water by the scoop-shaped tarsi of the short forelegs.

Backswimmers – Family Notonectidae

Most free-swimming aquatic animals have dark backs and light bellies, rendering them less visible against the light sky or dark depths. Backswimmers, true to their name, swim and float on their backs, and so their color pattern is reversed to provide the same protection against predators. Their fringed hind legs, much longer than their middle and front legs, are used as oars, and their keel-shaped back provides directional stability.

Backswimmers spend much time floating and resting at the surface of ponds and lakes, with head and body angled downward. They prey upon aquatic insects, terrestrial insects trapped in the surface film, and even small minnows and tadpoles. Members of this family trap bubbles of air in fringed pockets on the ventral surface, and can carry enough to survive up to six hours of inactivity underwater. Slightly longer than water boatmen (5–16mm long), backswimmers can also be distinguished from the former by their more streamlined shape, lack of dark cross-lines on the dorsal side, and lack of scoop-shaped tarsi on the forelegs.

Waterscorpions – Family Nepidae

Waterscorpions are easily recognized by their combination of scissor-like raptorial forelegs, sprawling middle and hind legs, and two very long abdominal appendages. When held together, these two filaments form a snorkel through which the insect breathes while clinging upside down to aquatic plants.

Most waterscorpions are slender and long – about 20–43mm in length – but those of the genera *Nepa* and *Curicta* resemble giant water bugs except for their tail-like breathing tube and lack of flattened hind legs. All are poor swimmers, and hence must remain motionless while awaiting the approach of unsuspecting prey.

Water Striders – Family Gerridae

Who among us does not have a childhood memory of idly staring at the edge of a pond or a quiet eddy in a stream and watching water striders skating about its surface with ease? They exploit the phenomenon known as surface tension, that property of water by which molecules at the surface cling more tenaciously to one another than they do elsewhere. This creates a film upon which light-weight objects can rest without breaking through. The same principle can be seen when water is

---

What you can do

Messing with Mother Nature

Have you ever been annoyed by the way in which impudent water striders taunt you, coming temptingly close, only to dash away before you can catch one? Here's how to get even! When you finally manage to capture one or more, place them in a container partly filled with clean, fresh water. Notice the dimples made by their feet as they depress the surface but do not break through.

Now, place a drop of dish detergent or light household oil on the water and watch your captives flounder. Both of these agents break the surface tension of water. You may wish to repeat this experiment with other substances to find out which have the same effect. Be sure to rescue your water striders before they drown, and discard the water away from the stream or pond.

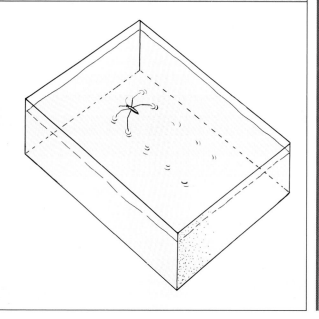

82

Above A small colony of the birch shieldbug (*Elasmostethus interstinctus*), from the family Acanthosomidae, contains one adult with fully developed wings, and four nymphs with developing wing buds.

Center In at the kill – water striders (*Gerris lacustris*) drink the blood of a dead insect (the large red damselfly, *Pyrrhosoma nymphula*) by piercing the hardened exoskeleton with their tough mouthparts. Water striders move rapidly over the surface film of the water with the aid of their long legs.

Left The brightly-colored *Lygaeus equestris* (family Lygaediae) is a common insect in southern Europe. It uses its warning colors as defense, and thus has no need to seek shelter as it sucks the juices from plants.

poured slowly into a glass until the level is higher than the lip of the glass itself. In the case of water striders, waterproof hairs on the undersides of their tarsi, combined with sprawling middle and hind legs that distribute their weight over a large area, enable them to walk on water.

Water striders prey upon the rain of insects that constantly fall onto any body of water in warmer months. Their struggle to escape the water sends ripples radiating across the surface, and the water striders use these as a form of homing signal. They use their raptorial front legs strictly for capturing prey, while darting across the water on their middle and hind legs. Water striders are generally slender and dark brown or black, their bodies measuring between 10mm and 25mm in length. This family includes the only true marine insects: certain species of the genus *Halobates*, found on warm subtropical seas.

Seed Bugs – Family Lygaeidae

Most members of this family, the second largest in the order, feed on seeds by injecting saliva and sucking out the partly digested contents. A few feed on other green plant tissues or sap, or prey upon other insects. They vary in appearance, with oval or elongated bodies from 3mm to 18mm long. Most are brownish, but some are boldly patterned with bright colors. They can be distinguished from members of most similar families by the presence of simple eyes and by the four or five prominent veins in the membrane of each forewing.

Leaf-footed Bugs – Family Coreidae

The odd name of leaf-footed bugs was derived from the unusual leaf-like expansions on the hind tibia of some species. Many also have enlarged hind femurs bearing prominent spines. These fairly large bugs, measuring from 7mm to 40mm long, can be differentiated from the seed bugs primarily by size and by the numerous parallel veins in the membrane of the forewing. Most leaf-footed bugs are plant eaters, and all can emit a foul odor when disturbed. Members of this family are also commonly known as squash bugs.

Stink Bugs – Family Pentatomidae

It takes little imagination to guess how these very common bugs got their name. In self-defense they will emit copious quantities of an offensive liquid from their thoracic glands to repel enemies. They

are generally between 6mm and 20mm in length and display a very prominent scutellum. The basal segment of their 5-segmented antennae is considerably thicker than the second. The many branching veins in the membrane of each forewing constitute another identifying feature. Most stink bugs feed on plant sap, but a few prey upon insect larvae.

Bed Bugs – Family Cimicidae

Though not a large family, bed bugs are significant in that they include a worldwide pest of humans, *Cimex lectularius*. They are often reddish-brown in color, measure less than 7mm in length, and are very flat and broadly oval. Their wings are reduced to tiny vestigial appendages that are barely visible. Bed bugs are nocturnal parasites that feed on the blood of mammals and birds, but they are not known to transmit any human diseases. By day, they hide in small cracks and crevices. Amazingly, adult bed bugs can survive up to 15 months without food.

Opposite top Typical of the family Coreidae, this leaf-footed bug (*Narnia inornata*) has enlarged hind tibiae, which help in the overall disguise of the insect on plants, and in rapid locomotion. Here, it is drinking juices from the fruit of the cactus *Pilosocereus sartorianus* in Mexico.

Opposite below The striking jester bug (*Graphosoma italicum*) – a European stink bug – has a matt surface to its striped body.

Above A mating pair of shield bugs (*Eurydema* sp.) from Corfu, Greece, display bright warning colors. They belong to the family of stink bugs (Pentatomidae) and emit deterrent odors if disturbed or handled. These common insects are often found in gardens.

What you can do

Collecting tips for true bugs

Capture techniques	**Killing methods**
hand collecting	ethyl acetate killing jar
aspirator	immersion in 70 per cent
sweep net	alcohol
beating tray	**Preservation**
light trap	**techniques**
	pin
	point

85

Plant Bugs – Family Miridae

Plant bugs constitute the largest family in the order Hemiptera, with some 1,700 North American species and roughly 6,000 species worldwide. Most feed on plant juices, and many are significant pests, but a few prey upon other insects. They are somewhat soft-bodied and often brightly colored. This family is marked by the presence of a *cuneus*, a distinct division of the wing along the anterior edge between the membrane and the basal section, called the *corium*. Close inspection will reveal a 4-segmented beak and 3-segmented tarsi. Simple eyes, or ocelli, are absent, and the head is shorter than the prothorax. Plant bugs are less than 10mm long.

Assassin Bugs – Family Reduviidae

Deserving of their ominous name, assassin bugs are named for the cool and calculating manner in which they stalk their quarry, deliberately raise a dagger-like beak into striking position, and then stab the victim, injecting a paralyzing saliva. Enzymes in the saliva break down the prey's internal tissues into a partially digested liquid, which the assassin bug then consumes much as one might drink a beverage through a straw. When not in use, the 3-segmented beak fits snugly into a central groove between the forelegs, again calling to mind the image of an assassin as he sheaths his weapon.

The femurs of an assassin bug's forelegs are enlarged to house the powerful muscles that hold

Below Nymphs of firebugs (*Pyrrhocoris apterus*) from southern Europe sun themselves on lime twigs. Firebugs are frequently sun-worshippers, or heliotropes. These gregarious insects derive safety in numbers and in pre-adult stage have not fully developed their wings.

the struggling prey until it is subdued. The bodies of assassin bugs may be either oval or elongated, and adults range from 12mm to 36mm in length. Their heads are always elongated, and are clearly marked by a deep groove running from one eye to the other.

Ambush Bugs – Family Phymatidae

Unlike assassin bugs, ambush bugs do not stalk their victims. Rather, they rely upon their cryptic coloration to lie in wait, usually on a flower, until a visitor seeking nectar or pollen ventures too close. Though rather small themselves (8–12mm), their stout build and enormous muscular front femurs enable them to take prey much larger than themselves, including butterflies, wasps, and bumblebees. Colored with various tones of green, yellow, and/or brown, ambush bugs typically have short, club-shaped antennae and a widely flared abdomen, extending well to the sides of their wings.

Lace Bugs – Family Tingidae

The identity of lace bugs is readily apparent from their elegantly sculptured forewings and expansive thorax. Due to the aforementioned features, these small bugs (3–6mm long) appear to have a rather unusual, rectangular shape. They are strictly plant feeders, and a species will often frequent only a single host species. Sap drips from the tiny wounds made by females as they lay eggs on the undersides of leaves, often resulting in formations like miniature stalactites as it dries.

Giant Water Bugs – Family Belostomatidae

The largest true bugs, 12–65mm in length, the giant water bugs' most obvious feature, besides size, is their powerful raptorial forelegs, which they use to grasp and hold prey as they thrust sucking mouthparts into it. They swim by means of flattened hind legs, returning to the surface regularly to breathe through two short, retractable tubes at the tip of the abdomen. Most of the larger species lay their eggs on aquatic plants, but some, particularly those of the genus *Belostoma*, cement the eggs to the back of the male, who carries them about until they hatch. The prey of giant water bugs includes other insects, tadpoles, salamanders, small fish, and even snails. You can study these water-dwelling behemoths by attracting them to lights at night.

What you can do

Building a micro aquarium

If you've been adventuresome enough to try close-up photography of insects and have achieved promising results, a new challenge awaits! While many terrestrial insects can be photographed in their natural habitats, this is virtually impossible where aquatic insects are concerned. To take satisfactory photos of these subjects, you will have to build some micro aquariums, which lend themselves well to indoor photography.

Use 5cm squares of slide mounting glass, available at photographic supply stores, and silicone bathtub caulk to make a cube that is open at the top. Glue the sides to the outside edge of the bottom section, instead of to the top, to allow room to insert a vertical glass slide into the finished aquarium to maneuver subjects into position. Use distilled water instead of pond water to ensure clarity.

Before shooting, remove all air bubbles from the sides with an artist's watercolor brush. Rinse the subject thoroughly in distilled water before transferring it to your aquarium. To achieve natural-looking photographs, make a background by swirling watercolors on a piece of paper to achieve a mixture of green, tan, and brown hues, simulating aquatic vegetation and debris.

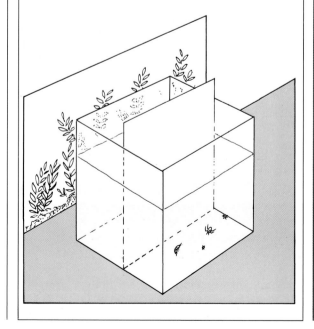

APHIDS, CICADAS, & HOPPERS
ORDER HOMOPTERA

SELECTED FAMILIES

Although entomologists once considered homopterans to be a suborder of the Order Hemiptera, the differences between the two groups have now been recognized as sufficiently significant for them to be placed in a separate order. Homoptera means "similar wing," reflecting the fact that their wings are uniformly membranous, in contrast to the unique wing design of true bugs. At rest, the wings of cicadas and hoppers slope in a roof-like manner over the body. Also different from those of true bugs are the mouthparts of homopterans, which, while again of the piercing-sucking variety and beak-like, arise far toward the posterior end of the head and can be projected down but never forward as they can among many hemipterans.

Members of this order feed exclusively on plants, but the appetites of individual species are fairly discriminating and vary tremendously. Cicadas and hoppers are vectors of some viral plant diseases, and much damage to agricultural crops and greenhouse plants is caused by the feeding of aphids and scale insects.

Cicadas and hoppers have large, razor-like ovipositors with which they cut slits and lay their eggs in the tissues of their favorite food plants. All homopterans undergo hemimetabolous development, also known as simple or incomplete metamorphosis. Nymphs resemble the adults except for their wings, which are reduced or absent.

There are about 45,000 species of homopterans worldwide, and 6,000 in North America. The larger species have distinctive features and are fairly easy to identify.

Cicadas – Family Cicadidae

Large, stocky insects, cicadas are best known for the shrill calls of the males, audible for up to a mile or more. Located on the sides of the first abdominal segment are membranous areas of the cuticle, called tymbals, which are alternately buckled and restored by the contraction and relaxation of a large muscle. Each movement of the tymbals produces a click or pulse, which in a long series constitute the cicada's call, which is a hallmark of the warm, dry days of summer. A complex series of resonating air sacs present in the abdomen greatly amplifies the call.

Cicadas are dark and fairly large – 16–60mm long – with prominently veined, membranous wings, the front pair of which are about twice as long as the hind pair. Females lay their eggs in slits made in twigs of trees and shrubs, the tips of which usually die as a result. Upon emerging, nymphs fall to the ground, where they burrow under the soil with well-developed fore legs and feed on the xylem sap of plant roots. Because of the low nutrient content of this food, development

---

What you can do

Cicada hunting

Excluding the years when large broods of periodic cicadas emerge simultaneously, these robust insects can be surprisingly hard to spot because of their preference for tall shade trees. Your best chance of finding them is to follow the male's raucous buzz, which swells in magnitude and then tapers off toward the end. If you cannot spy the adults, then look for their shed nymphal skins, which are hollow replicas, down to the smallest detail. You will find these still clinging to the bark of tree trunks where they were abandoned.

If you find these, there are almost certain to be burrows at the base of the tree, from which the nymphs emerged. These burrows are roughly 15mm in diameter and show no signs of excavation from the surface, such as loose dirt around the perimeter.

Above left This Australian cicada is waiting for its new wings to harden. Its body is soft also, but it will take under an hour to dry and darken.

Above This male cicada of the *Macrotristria* genus from Australia is ready to fly, since its transparent wings have hardened and its colors are fully developed.

Left Following a mass emergence of cicadas, they may all line up to dry their wings; here, a *Platypedia* species rests on a scrub oak twig in Colorado.

89

is slow, and the nymphs will spend the next 4 to 17 years underground, depending upon the species. Occasionally, enormous broods will emerge in the same year, leaving thousands of shed skins still clinging to the individual trees that they climbed for their final molt.

Treehoppers – Family Membracidae

Treehoppers are notable for their large pronotums that project far forward over the head, or backward over the abdomen, or both, often in bizarre protective configurations that make the insect appear as a thorn or another part of the plant upon which it feeds. They are small jumping insects, about 5mm to 12mm in length, but able to clear a meter or more in a single bound. Some are boldly marked with bright colors; in some instances, females exhibit duller colors than males, whose long-term survival is less important to reproductive success.

Treehoppers feed on the xylem sap of plants, usually trees and shrubs. Copious quantities of this nutrient-poor liquid must be ingested, and is voided as "honeydew," a sweet anal secretion relished by ants, who tend small "herds" of treehoppers much as a shepherd tends livestock. The ants get the honeydew they so desire, while the treehoppers are protected from such predators as spiders.

Spittlebugs – Family Cercopidae

Also known as froghoppers for their frog-like appearance, spittlebugs get their name from the frothy white masses manufactured by the nymphs as a moist refuge. Produced as air is blown into an anal secretion, these bubbly accumulations are commonly seen on the grasses and herbaceous plants upon which the nymphs usually feed, and sometimes on trees and shrubs as well.

Spittlebugs are small, between 4mm and 13mm in length. Grayish or brownish adults leap about freely and are easily distinguished from the similar leafhoppers by having one or two prominent spines on each hind tibia, rather than the two rows of spines that are found in the same location on leafhoppers.

Below Here disguised as plant thorns, these thornbugs (*Umbronia spinosa*), from the Peruvian rainforest, demonstrate the amazing diversity of body shape found in this species.

Below right The familiar "snake-spit" on plants is actually a protective bubbly secretion produced by the nymphs of the spittlebug, as by this member of the *Philaenus* genus.

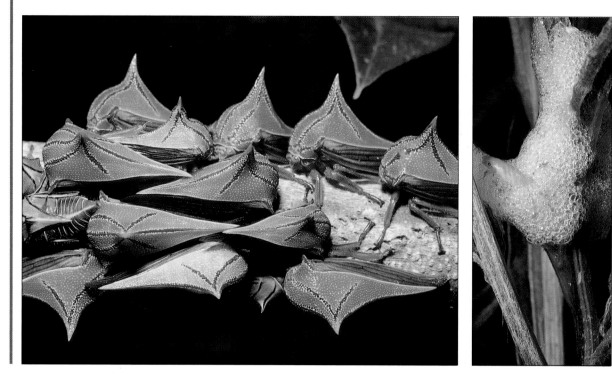

Top The red and black froghopper (*Cercopsis vulnerata*) is a common European species; usually found on plants, it will jump or fall to the ground if disturbed.

Above Each only a few millimeters long, gregarious whiteflies (*Aleyrodes brassicae*) are a pest in greenhouses, breeding on the undersides of leaves.

Leafhoppers – Family Cicadellidae

Though small in size (2–15mm in length), leafhoppers constitute the largest family in the Order Homoptera. Many are brightly colored, and the two rows of spines on each hind tibia are a unique feature. Leafhoppers are found on all types of vascular plants, but are most common in meadows in other grassy areas. Most leafhoppers feed on phloem sap of plants, which is rich in organic nutrients. Among those that utilize plants' xylem sap, some are nicknamed "sharpshooters," because of the frequent and forceful method in which they expel honeydew. This family includes many plant pests; these damage the plants by their feeding and also transmit viral diseases.

Whiteflies – Family Aleyrodidae

Tiny insects, measuring only 1mm to 3mm in length, whiteflies are common outdoors in tropical and subtropical areas, but they have become pests of greenhouses and indoor plants everywhere, appearing as tiny white flecks. The wings and bodies of whiteflies are covered with a white, powdery substance. In all but the first instar, nymphs are immobile, attached to the plant by a fringe of waxy white filaments and resembling scale insects. The wings of adults are disproportionately large and fold horizontally over the body at rest.

Aphids – Family Aphididae

The sight of pear-shaped aphids, colored bright green, red, brown, or various pastel tints, is enough to strike terror into the hearts of gardeners. What makes aphids so destructive is their tremendous reproductive capacity. Unlike most other insects, which reproduce sexually, aphids can reproduce by *parthenogenesis*, a process in which the eggs develop without fertilization by a male and hatch into exact genetic replicas of the mother. In this fashion, billions of offspring can originate from just one female aphid, causing massive infestations within a short period of time. Their sheer numbers weaken and kill plants by consuming large quantities of sap and causing them to wilt and turn yellow.

A rather complex life cycle accounts for aphids' existence. Eggs that overwinter hatch into wingless females in the spring. These reproduce by parthenogenesis and are among the few truly viviparous insects, nourishing the embryos from

91

Above A group of aphids (*Hyalopterus pruni*), of different ages and colors, are here feeding and breeding on the stem of a reed. None of these aphids is winged, and their colors reflect their different ages and the juices they have been drinking.

Right A female nettle aphid (*Microlophium carnosum*) gives birth while drinking juices from the plant. The spines and hairs on the nettle do not appear to deter the winged adult aphids.

What you can do

Aphid ranching

It's not the Wild West, but aphid ranching will allow you to observe the mutually beneficial relationship between ants and aphids. Follow the procedure given on pages 142–3 for setting up an ant farm. Next, in late spring or summer, go hunting for a small plant bearing an aphid colony and carefully transplant this into a pot, or set several potted plants, such as roses, outdoors as "bait." (Be careful not to remove any rare plant from its natural habitat, however.) Place the entire potted plant either into an aquarium, sealing this with fine mesh and a large rubber band, or into a tall, clear plastic or cellophane bag, using a tall wooden dowel rod for support. Add the ant farm, opened to allow them access to the aphids, seal both inside, and place the ranch where the plant can get its natural amount of sunlight. Within a short time, the ants will begin to "milk" the aphids, stroking them with their antennae in a way that stimulates them to release their honeydew. When the ants have come to treat the aphids as "theirs," introduce a natural aphid predator, such as a ladybird beetle, and watch the ants' reaction.

Above Aphid-farming is a well-known aspect of the social life of ants, and here *Myrmica ruginodis* attend aphids of *Aphis viburni* on the stems of a guelder rose. It is in the interest of ants to look after aphids since, on stimulation, the aphids produce a sweet liquid on which the ants gorge themselves.

their own tissues and giving birth rather than laying eggs. After several generations, winged females appear; these migrate in a swarm to another species of plant, and continue to reproduce asexually. Late in the growing season, they return to the original host species of plant, where they produce a generation of males and females which mate and deposit another overwintering batch of eggs.

Aphids are sought out by many species of ants for the sweet anal secretions, or "honeydew," they produce. The ants zealously tend and protect them as if they were so many tiny cattle, even going so far as to carry them to the more nutritious parts of the plant.

Scale Insects – Superfamily Coccidea

This group includes several families characterized by their habit of secreting a waxy, scale-like covering on plants in order to conceal themselves or their eggs. These scaly shelters vary greatly in appearance, texture, and composition. They have been economically important, not only as plant pests, but also as the raw material from which certain red dyes, called cochineal dyes, were extracted and used to color cosmetics, medicines, and beverages. They were also the source of *lac*, from which shellac was formerly made.

Nymphs and adult females are immobile under their protective scales, while adult males have one pair of wings. A needle-like appendage at the tip of the abdomen and their lack of mouthparts sets adult male scale insects apart from gnats, which are in the Order Diptera. Identification of scale insect families is based mostly on microscopic features of the females.

What you can do

Collecting techniques
by hand or forceps
aspirator
sweep net
beating tray
light trap

Killing methods
ethyl acetate killing jar
immersion in 70% alcohol

Preservation techniques
pin through right side of scutellum
point
70% alcohol vials

BEETLES
ORDER COLEOPTERA

SELECTED FAMILIES

Coleoptera means "sheath wings," a reference to the armored forewings, called *elytra*, which are either hardened or leathery and, when folded, meet in a straight line down the middle of the back, protecting the membranous hind wings that are used for flight. In flight, the elytra are held perpendicular to the body and function as airfoils, providing added lift. Many species have parallel lines or grooves called *striae* running the length of each elytron. Although nearly all beetles can fly, most do so only to cover short distances or to reach low vegetation. The rest of their time is spent either crawling on or near the ground or on vegetation or swimming.

All beetles have chewing mouthparts with well-developed mandibles that, depending on the species, are variously adapted to partake of a wide variety of foods. Most beetles are either herbivorous or scavengers, but there are also many predators among them. Many of the plant-eaters are destructive pests of crops and forests and in large numbers can cause significant damage, either directly or by transmitting diseases. Other problem beetles infest stored food, clothing, carpets, or museum specimens.

Beetles have two prominent compound eyes, and antennae of various shapes and sizes that arise between the compound eyes. The pronotum is usually quite prominent, but the other two thoracic segments, as well as most of the abdomen, are typically hidden under the elytra when viewed from the dorsal side. The metamorphosis of beetles is complete – in other words, they are holometabolous – and the larvae do not resemble the adults.

Aside from the preceding information, it is very difficult to generalize about beetles. There are more than 300,000 known species of beetles in the world, a number approximately equal to that of known plant species. They constitute about 40 per cent of all known insects, and about one-third of all known animal species. Members of this order range in size from 0.25mm to 200mm in length, up to 75mm in width, and they occur in many shapes and in all colors, although most are dark. You can find them in every conceivable habitat, except for salt water and polar ice caps. It could well be argued that beetles are the most successful organisms on earth.

Tiger Beetles – Family Cincindelidae
As their name suggests, tiger beetles are fierce predators, especially upon other insects. They are long-legged insects that run and fly swiftly, but unlike the tiger, which relies upon stealth, tiger beetles are more likely to use their speed to run down prey. Typically, they fly close to the ground for short distances.

Tiger beetles are a colorful group, sometimes adorned with an iridescent or metallic sheen, and

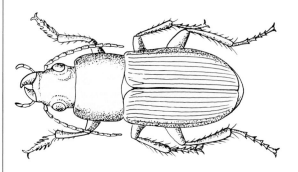

Above There are more than 300,000 known species of beetles in the world, ranging in size from 0.25mm to 200mm in length. The Latin name of their order refers to the armored forewings which meet in a straight line over their back when folded. All beetles have chewing mouthparts with well-developed mandibles, variously adapted for a wide range of food.

Above The bombardier beetle (*Brachinus crepitans*) bombards its predators with an irritating and foul-smelling aerosol, which it sprays from its abdomen when disturbed.

Left Lively tiger beetles (*Cincindela campestris*) live in sandy areas and fly off if approached too quickly. Here, a male attempts to mate; occasionally, other competing males will arrive and ride piggyback. The bright metallic colors are typical.

ranging in length from 6mm to 40mm. They are easily identified by their wide head and prominent sickle-shaped mandibles that cross in front of the head. Most are active during daylight and are fond of open, sunlit, sandy areas like beaches and paths. Handle them very carefully to avoid a painful bite.

Ground Beetles – Family Carabidae

Like the tiger beetles, with which they are sometimes lumped into a single family, most ground beetles actively pursue prey, but they are nocturnal predators. During the day, you will usually find them in moist places, under rocks, logs, or leaf litter, for example. Ground beetles are from 3mm to 36mm long, and long-legged; most are shiny and dark, although some are quite colorful.

The members of one particular group, the genus *Brachinus*, are known as bombardier beetles because of their unique defense mechanism. Two relatively benign chemicals are stored in a special abdominal chamber. When the beetle is disturbed, these chemicals are injected into another chamber, where they are acted upon by an enzyme to produce a violent reaction, releasing oxygen, water, noxious chemicals called quinones, and considerable heat. This mixture, which approaches 212°F (100°C), is released as an explosive puff of irritating gas that can be aimed by the beetle in a 360 degree radius with remarkable accuracy. The heat, irritation, and an audible report as the spray is released are enough to deter most would-be predators, and the beetle can reload and fire in rapid succession – up to a dozen times if necessary – before its reservoir is temporarily exhausted.

Predacious Diving Beetles – Family Dytiscidae

These fairly large aquatic beetles are common in most freshwater habitats. Their streamlined oval bodies are from 5mm to 70mm long and are usually dark green, yellowish-brown, or brownish-black in color. Predacious diving beetles swim by moving their hind legs in unison like oars. In this, they are unlike water scavenger beetles, with which they may be confused but which move their hind legs alternately. They must visit the surface periodically to renew their air supply, which is stored as a bubble in a special chamber under the elytra. Unlike water scavenger beetles, which break the surface film head first, predacious diving beetles come to the surface tail first.

They overwinter as adults, frequently migrating from one body of water to another in spring or fall. They fly well, and are attracted to lights at night.

Below These whirligig beetles (*Gyrinus natatior*) are about 6mm long, and are able to race over the water film in short bursts at a speed of about 40in per second.

Right The great diving beetle (*Dytiscus marginalis*) is found throughout Europe. This specimen is breathing at the water surface.

Whirligig Beetles – Family Gyrinidae

Early spring and middle to late summer are the times when you are most likely to see these shiny black or metallic dark green beetles darting in a seemingly aimless fashion across the surface film of fresh water. Whirligig beetles tend to congregate in "herds," floating serenely or gyrating in whirling patterns on calm areas of ponds, lakes, marshes, and stream eddies. They can dive underwater, and frequently do so to escape danger. The oval, flat adults are 3–15mm long, and are easy to recognize by their compound eyes, which are divided into upper and lower sections which can see both above and below the surface of the water simultaneously. Scoop-shaped segments on their short, clubbed antennae enable them to read vibrations on the surface film of the water, by which they avoid obstacles and locate prey.

Hister Beetles – Family Histeridae

These small, hard-bodied beetles are normally shiny black, but may also be bronze or green. They are less than 10mm long and their shortened elytra leave the last one or two abdominal segments exposed. Hister beetles generally live near decaying organic matter, such as carrion or dung, where they prey on other small insects. A few species live under loose bark or in termite or ant colonies.

Rove Beetles – Family Staphylinidae

The very short elytra of rove beetles, which covers only the first few abdominal segments, are most conspicuous. They are generally between 2mm and 20mm long, often with slender, ant-like bodies. Rove beetles run rapidly, with the tips of the abdomen curled up and forward like a scorpion's stinger. Despite their short elytra, they are among the best fliers of all beetles. Most are predators, and they are often found on flowers, mushrooms, under bark, or in leaf litter.

Below Rove beetles, of the Staphylinidae family, are named after their behavior of roving over the ground with the tip of the abdomen lifted upward looking for prey. One of the hind wings of this *Ontholestes tessellatus* is just visible, beneath its short elytra.

97

Carrion Beetles – Family Silphidae

Insects such as carrion beetles provide one of nature's waste removal services, while at the same time enhancing the soil with their own nutrient-rich waste. Some carrion beetles excavate under the dead animal and bury it after laying eggs on it, so their offspring won't have to compete with fly larvae for food. Adults actually feed primarily on maggots on the carcass.

Carrion beetles are the largest such scavengers, ranging in length from 1.5mm to 40mm, and are usually black, with bright orange, red, or yellow markings. Somewhat flattened, with club-shaped antennae, they are easily attracted to traps baited with small dead animals. At least one species is attracted to dead animals suspended in the air, and most show a preference for certain types of animal, such as birds, snakes, mammals, fish, and so on. When threatened, some may feign death, while others emit an unpleasant odor or a foul-tasting secretion, a defense advertised by their bold markings.

Stag Beetles – Family Lucanidae

The name of this family was derived from the massive, branching mandibles on males of its larger species, which resemble the antlers of a stag and are used in similar fashion for defense and to resolve disputes over females. The beetles can use these to inflict a painful pinch if mishandled, and so are also known as pinching bugs. They range

Above The stag beetle (*Lucanus cervus*) spends up to three years as a larva eating rotting wood in old logs, and then emerges from an intermediate pupal stage into an adult about 50mm long.

What you can do

A-maze-ing beetles

If you are fortunate enough to find a large scarab beetle, especially one of the dung or carrion feeders, you can actually see them smell. Construct a small maze out of sturdy poster board or light plywood (the walls need only be centimeters high) and cover it with a clear plastic. Place a piece of dung or decaying meat as bait at one end of the maze and introduce the beetle at the other end, sealing the entrance to force the beetle into the maze. As it moves along, trying to locate the bait, you will see the plate-like processes on the tip of its antennae unfurling from tight balls as they fan out to "sniff" the air, exposing the full area of each to intercept volatile particles.

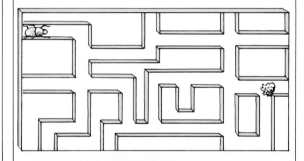

Above Carrion beetles, such as this one – *Nicrophorus vespillo*, of the Silphidae family – offer a valuable service breaking up and recycling dead animals and decaying vegetation. This specimen shows enlarged antennae, tibial spurs, and golden hairs on its body.

from 8mm to 60mm in length and have *lamellate* antennae, wherein the last 3 or 4 segments form a comb-like process to one side. Stag beetles are common in old woodlands, especially in decaying logs, on which the larvae feed. Despite their formidable mouthparts, adults are reported to feed on aphid honeydew and the sap of trees and leaves. Some species are attracted to lights during summer.

Scarab Beetles – Family Scarabaeidae

Scarab beetles are mostly nocturnal scavengers, playing an important role in recycling carrion, dung, and decomposing vegetation. Their front fore legs, or tibia, are wide, flattened, and sometimes toothed to serve as shovels or rakes for excavation. Members of this family are stocky, with large heads and pronotums and a highly convex body 5–60mm long; many are adorned with striking metallic colors. They have lamellate antennae tipped with comb-like processes that can be rolled into a tight ball or unfurled. Light traps and traps baited with carrion, dung, or a fermented molasses-yeast-water mixture are effective collecting tools for scarab beetles. The Scarabaeidae includes the well-known June, Japanese, and rhinoceros beetles.

Soldier Beetles – Family Cantharidae

Soldier beetles are similar to fireflies, but they cannot produce light and lack the extended pronotum that covers the head of a firefly. The pronotum and elytra are usually red, yellow, or orange. Downy hair usually covers the elytra. Most are predatory, but a few visit flowers to feed on

Below Beetles are the most diverse group of insects on earth, and stag and rhinoceros beetles, such as this rhinoceros beetle from Mexico (*Golofa pizarro*), are among the most spectacular. The enlarged rhino spurs are for display, intimidation, and courtship fights.

pollen and nectar, and some are omnivorous. Soldier beetles are common in tall grass.

Net-winged Beetles – Family Lycidae

Members of this family have elegantly-sculptured elytra, with a delicate network of veins, interrupted by prominent, parallel ridges that run the length of each forewing. The elytra are widest toward the rear, giving the insect a pear-shaped appearance, and this effect is enhanced by the rounded pronotum, which conceals most of the head. Net-winged beetles are generally 5–19mm long, and red, yellow, or orange, often with black markings. You will usually find them during the day in densely wooded or moist areas, on flowers, foliage, or tree trunks. They are best collected by sweeping or beating the vegetation in such areas.

Checkered Beetles – Family Cleridae

A wide head with bulging eyes and a narrow, elongated body bristling with erect hairs are clues to the identity of members of the Family Cleridae. The pronotum is narrower than the elytra, and often nearly cylindrical in shape. Many sport bold markings of orange, yellow, red, or blue, sometimes in a checkered pattern. Some checkered beetles feed on pollen and are found on flowers and foliage, while many others prey upon wood-boring insects and are most commonly found on the trunks of dead or dying trees. They are rather small, usually between 3mm and 12mm in length.

Metallic Wood-Boring Beetles – Family Buprestidae

This large family is named for the metallic green, blue, copper, bronze, or black sheen of adults, which is especially noticeable on the ventral side of the body and on the dorsal side of the abdomen. The parallel sides of the body and the pointed elytra give metallic wood boring beetles somewhat of a bullet shape. Adults are attracted to dead or freshly-cut wood and tree trunks, branches, leaves, and flowers in sunlit locations, and so are especially common on the southern edge of forests in the northern hemisphere. The winding paths made in the sapwood under tree bark by their larvae, known as flathead wood borers, often kill or severely damage trees.

Click Beetles – Family Elateridae

As defenses go, that of click beetles is a surprise –

Above This metallic wood-borer, Capnodis tenebrionis, is now very rare in Europe. The larva spends two years underground eating the roots of plants; the adult is found in the spring on various plants.

Above and **right** The tiny elm bark beetle (Scolytus sp.), which is only 2mm long, carries the fungus that killed million of elms in the United States and Europe in the 1970s. Larvae from eggs laid below the bark radiate out to form galleries as the larvae become bigger. Many trees suffer from their own species of bark beetles.

quite literally. The oversized prothorax of a click beetle is loosely hinged to the rest of its body. Should a bird, reptile, or other animal seize it, the click beetle quickly arches its body at this special joint and, with a sudden flex in the opposite direction, snaps a spine-like extension of the prosternum into a snugly-fitting groove on the mesosternum. The resulting audible and tangible "click" usually startles the would-be predator into dropping the beetle which, it it lands on its back, clicks again and again, catapulting itself into the air until it lands right side up, to make its escape. Smaller click beetles are more adept at this than are their larger cousins, which may be up to 50mm long. Click beetles are easily recognized by their elongated bullet shape and the unusually large pronotum, which has pointed posterior corners.

Fireflies – Family Lampyridae

There are few spectacles in the insect world more dramatic than the sight of a meadow full of twinkling fireflies on a moonless summer night. Also known as lightning bugs, they produce their flashing luminescence in a special organ near the tip of the abdomen, where an enzyme, luciferase,

Below Wireworms, which are very common in soil and old timbers, are the larvae of click beetles. This colorful click beetle belongs to the *Alaus* genus from Kenya, and is looking for a suitable place to lay eggs. If disturbed, it immediately jumps, giving an audible click sound.

Bottom Sprung for action, this click beetle (*Chalcolepidius porcatus*) holds the extension from its pronotum into the notch from which it can be released to right itself when upside down or to startle would-be predators.

acts upon another substance, luciferin; not all species, however, are capable of producing light. The color, frequency, and sequence of the flashes are unique to individual species, and help males and females to locate and recognize one another. Most of the fireflies that we see are males, which fly around open areas while emitting their specific signal. Upon recognizing the correct code, a female perched on or near the ground will respond, and the male quickly homes in on her beacon. The females of a few species mimic the signals of other species and devour the males.

Although there are about 2,000 species of this family in the world, fireflies are best represented in the Americas and Asia, the best example in Western Europe is the glowworm. Fireflies are medium-sized insects, from 5mm to 20mm in length, with rather soft elytra and a forward extension of the pronotum that shields the head. Females may be short-winged or wingless. Those that emit light are obviously nocturnal, and each species flashes actively only during a limited period of the night. Much of the rest of their time is spent clinging to foliage, tree trunks, branches, or in moist places under bark or decaying vegetation.

What you can do

Collecting tips
Capture techniques
by hand or forceps
aspirator
sweep net
aerial net
aquatic net
beating tray
bait trap
light trap
Berlese funnel

Killing methods
ethyl acetate killing jar
immersion in 70% alcohol

Preservation techniques
pin through right elytron
point
vial of 70% alcohol

Note Certain large members of the ground beetle and scarab beetle family develop greasy deposits on their surface after pinning. Remove the labels from the pin and degrease the beetle by immersing it repeatedly in a small amount of organic solvent, such as alcohol or acetone, replacing the solvent until it no longer becomes discolored.

Ladybird Beetles – Family Coccinellidae

Oft portrayed in children's books and nursery rhymes, the amicable ladybird beetle is one of our greatest insect allies. Also known as ladybugs and lady beetles, these diminutive – between 1mm and 10mm long – dome-shaped insects and their larvae are voracious predators upon such insect pests as aphids, scale insects, mealybugs, and mites. So effective are they that many garden supply companies sell them as biological control agents. Their round, convex, and colorful forms and their abundance make them one of our most familiar insects, despite their small stature. Most are red, orange, or yellow with black spots, but this color pattern is reversed in some species, and others are spotless. They frequently overwinter as adults in huge swarms under bark and leaf litter, a practice that seems to produce and conserve warmth.

Above Glowworms (*Lampyris noctiluca*) belong to the same sub-family of beetles as fireflies, and have a wide distribution in Europe. The light is produced only by females, as a means of attracting the winged males.

Above Ladybirds of the 11-spot species (*Coccinella 11-punctata*) gather together during their late summer sleep and during hibernation over winter. They often choose to squeeze into tree stumps and cracks in wood, where they remain dormant for several months. They may seek their winter refuges at the tops of mountains, and can be found gathered together under stones.

103

Darkling Beetles – Family Tenebrionidae

A heavily striated or bumpy elytra is the hallmark of these mostly dull black or brown beetles, and close inspection will reveal their compound eyes to be kidney-shaped or notched, rather than round. Most are nocturnal scavengers, found under rocks, debris, or decaying wood, but some species infest and damage stored food, clothing, rugs, and museum specimens. They can also damage insect collections, a good reason to keep yours in a sealed case! Darkling beetles are slow-moving, small to medium in size (2–35mm long), with variable body shapes.

Long-horned Beetles – Family Cerambycidae

Long-horned beetles are named for their long, swept-back antennae, which, combined with their often brilliant colors, make them popular with collectors. The vast majority of species in this family have antennae that are at least half as long as their elongated, cylindrical bodies, and may be up to three times the body length, which itself measures 6–75mm. The larvae tunnel through the wood of dead, dying, and occasionally living trees to feed, then bore into the bark to pupate. Most adults feed either on the pollen, nectar, and stamens of flowers, or on foliage, fruit, sap, roots, or fungi.

Though they are the bane of the timber industry because of the defects their larvae cause in recently-felled trees, long-horned beetles accelerate the decomposition of dead trees, and are therefore a vital element in forest ecosystems.

Leaf Beetles – Family Chrysomelidae

True to their name, leaf beetles are herbivorous. Adults feed on flowers and foliage, but most larvae eat leaves or bore into roots and stems. The larvae of some species are leaf miners, tunneling between the outer layers of leaves and feeding on the spongy tissue, which often results in leaf deformities. They are more common in weedy, open areas and on bushes than on trees.

Though common and often brightly colored, leaf beetles can be somewhat difficult to identify because of their varying body shapes. Some resemble long-horned beetles, though their antennae are nearly always less than half as long as their body. Others look much like ladybird beetles, and these can be much more difficult to separate, sometimes requiring scrutiny of the tarsal

---

What you can do

Observing metamorphosis

One species of darkling beetle, *Tenebrio molitor*, is a particularly good specimen with which to study the stages of complete metamorphosis, also known as holometabolous development. Their larvae, known as yellow mealworms, are sold in pet stores as food for fish, reptiles, and amphibians, and so are readily available. An old goldfish bowl or large wide-mouthed jar makes an excellent home for breeding mealworms. Place a layer of bran or oatmeal, about 1cm deep, in the bottom of the jar, followed by an equal layer of dried bread or cookie crumbs mixed with flour. Top this with another layer of bran or oatmeal, and introduce several dozen mealworms. They will need a small amount of moisture, which can be supplied with a slice of apple or carrot, replaced daily. Add a piece of crumpled paper towel or tissue for the nocturnal adults to hide under, and to ensure adequate ventilation secure a muslin or similar cover over the top of the jar with a rubber band. With this simple set-up, you can raise many generations of the beetles and make detailed observations of their life history.

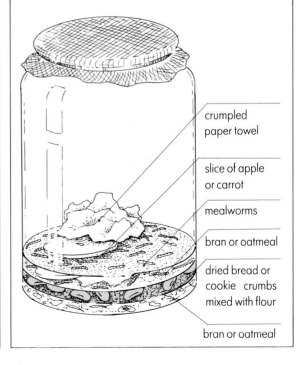

crumpled paper towel

slice of apple or carrot

mealworms

bran or oatmeal

dried bread or cookie crumbs mixed with flour

bran or oatmeal

segments, which number three in ladybird beetles and four in leaf beetles.

Weevils – Family Curculionidae

Most weevils have a prominent, down-turned snout, protruding forward from the head, at the end of which chewing mouthparts are located. Most species have elbowed, clubbed antennae, extending laterally from midway on the snout, which often has a groove to accommodate the long first antennae segments. Ranging in length from 1mm to 40mm, weevils are the largest family in the animal kingdom, with more than 40,000 species worldwide. All are plant eaters, and many are significant agricultural pests.

Right Long-horned beetles (family Cerambycidae) are named after their antennae, which are often longer than their bodies; this one, *Penthea pardalis*, is from a eucalyptus forest in Queensland, Australia.
Below left The remarkable protrusion of this weevil (*Curculio glandium*) is typical of this group of beetles (also known as snout beetles). The sensory pair of antennae are clearly seen.
Below right The Colorado potato beetle (*Leptinotarsa decemlineata*) is a native of the Americas, though it has now been introduced to Europe where it continues to be a pest on potatoes. Other varieties are yellow and black.

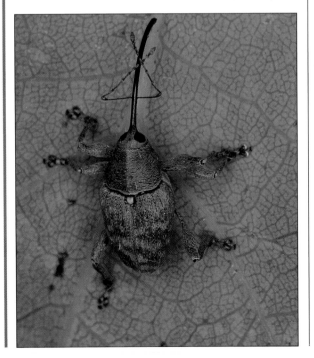

KEY TO COMMON BEETLE FAMILIES

Each step in the key asks a question about the specimen, usually answered by "yes" or "no," which then directs you to another part of the key or reveals the family name.

1 Does the front of the head extend into a long and narrow or broad and short snout, or is the body cylindrical and the antennae elbowed and club-shaped?
a. Yes .... Go to #74
b. No .... Go to #2

2 Is the insect aquatic, with hind coxae flattened into large plates?
a. Yes .... Haliplidae
b. No .... Go to #3

3 Is the posterior edge of the first abdominal sternite interrupted by the hind coxae?
a. Yes .... Go to #4
b. No .... Go to #8

4 Are the hind legs flattened and hairy?
a. Yes .... Go to #5
b. No .... Go to #6

5 Are the antennae long?
a. Yes .... Gyrinidae
b. No .... Dytiscidae

6 Is there a groove crossing the metasternum in front of the hind legs?
a. Yes .... Go to #7
b. No .... Rhysodidae

7 Is the head, including the compound eyes, as wide or wider than the pronotum?
a. Yes .... Cincindelidae
b. No .... Carabidae

8 Are the tarsi of the front and middle legs 5-segmented and the hind tarsi 4-segmented?
a. Yes .... Go to #49
b. No .... Go to #9

9 Are all tarsi 5-segmented?
a. Yes .... Go to #11
b. No .... Go to #10

10 Are all tarsi 3-segmented?
a. Yes .... Go to #72
b. No .... Go to #62

11 Do the antennae have clubbed, comblike projections?
a. Yes .... Go to #12
b. No .... Go to #18

12 Do the antennae have many comblike projections?
a. Yes .... Go to #13
b. No .... Go to #14

13 Is the ventral side hairy?
a. Yes .... Scarabaeidae
b. No .... Rhipiceridae

14 Can the last 3 or 4 plates of the clublike antennae be closed into a ball?
a. Yes .... Scarabaeidae
b. No .... Go to #15

15 Is the last tarsal segment much longer than the others?
a. Yes .... Go to #16
b. No .... Go to #17

16 Is the insect aquatic?
a. Yes .... Dryopidae
b. No .... Silphidae

17 Are the elytra marked by conspicuous longitudinal grooves?
a. Yes .... Passalidae
b. No .... Lucanidae

18 Are the hind legs curved and flattened for swimming?
a. Yes .... Hydrophilidae
b. No .... Go to #19

19 Do the elytra cover less than half the length of the abdomen?
a. Yes .... Staphylinidae
b. No .... Go to #20

20 Do the elytra cover all abdominal segments?
a. Yes .... Go to #21
b. No .... Scaphidiidae

21 Are fewer than seven ventral segments of the abdomen visible?
a. Yes .... Go to #24
b. No .... Go to #22

22 Do the middle coxae touch or almost touch each other?
a. Yes .... Go to #23
b. No .... Lycidae

23 Is the head concealed under the pronotum?
a. Yes .... Lampyridae
b. No .... Cantharidae

24 Are six ventral segments of the abdomen visible?
a. Yes .... Go to #25
b. No .... Go to #27

25 Is the body flattened and oval?
a. Yes .... Silphidae
b. No .... Go to #26

26 Are the elytra wedge-shaped?
a. Yes .... Melyridae
b. No .... Cleridae

27 Are the antennae both club-shaped and bent in an elbow?
a. Yes .... Histeridae
b. No .... Go to #28

28 Is the body dark orange or light brown and densely covered with yellow, gray, or white hair?
a. Yes .... Byturidae
b. No .... Go to #29

29 Is more than half of the head concealed under the pronotum when viewed dorsally?
a. Yes .... Go to #30
b. No .... Go to #37

30 Are the antennae at least half as long as the body and filiform (threadlike) in shape?
a. Yes .... Go to #31
b. No .... Go to #32

31 Do any of the antenna segments have projections at the base?
a. Yes .... Ptilodactylidae
b. No .... Ptinidae

32 Is the body elongated, narrowly oval, or cylindrical?
a. Yes .... Go to #33
b. No .... Go to #36

33 Are most antennae segments broader than they are long?
a. Yes .... Dryopidae
b. No .... Go to #34

continued

Above Beetles, with over 300,000 species, are the most successful order of organisms to inhabit the earth. One reason for this is their varied methods of defense. These blister beetles (*Tetraonyx frontalis*) from Mexico utilize bright colors to advertise the toxic chemical cantharidin in their tissues, which can cause blistering of the skin.

Continued from p107

34 Are the pronotum and elytra wrinkled in appearance, with curved ridges alternating with rows of puncture-like marks?
a. Yes .... Elmidae
b. No .... Go to #35

35 Is the pronotum armed with knobs, hooks, or teeth near the front?
a. Yes .... Bostrichidae
b. No .... Anobiidae

36 Are the last three or more antennae segments longer than the others?
a. Yes .... Anobiidae
b. No .... Helodidae

37 Do each of the tarsi have one or more lobes on the ventral surface?
a. Yes .... Go to #38
b. No .... Go to #41

38 Is the fourth tarsal segment the only one with a ventral lobe?
a. Yes .... Helodidae
b. No .... Go to #39

39 Do the first, second, third, and fourth tarsal segments all have ventral lobes?
a. Yes .... Cleridae
b. No .... Go to #40

40 Are the posterior corners of the pronotum distinctly pointed?
a. Yes .... Elateridae
b. No .... Dascillidae

41 Are the front coxae cone-shaped and prominent?
a. Yes .... Go to #42
b. No .... Go to #43

42 Is the beetle aquatic, with serrate antennae?
a. Yes .... Psephenidae
b. No .... Dermestidae

43 Are the front coxae at right angles to the anterior-posterior axis of the body?
a. Yes .... Go to #44
b. No .... Go to #45

44 Is there a gap between the posterior edge of the pronotum and the base of the elytra?
a. Yes .... Trogositidae
b. No .... Nitidulidae

45 Does a spine extend between the front coxae from the posterior edge of the prosternum?
a. Yes .... Buprestidae
b. No .... Go to #46

46 Are the last two or three segments of each antenna significantly larger than the others, and does the head narrow behind the eyes?
a. Yes .... Lycidae
b. No .... Go to #47

47 Are the cavities from which the coxae protrude touched by the lateral sclerites of the mesothorax?
a. Yes .... Cucujidae
b. No .... Go to #48

48 Is the pronotum much wider than the head, with the anterior corners somewhat pointed?
a. Yes .... Erotylidae
b. No .... Languriidae

49 Are the cavities from which the front coxae protrude closed behind the lateral sclerites of the pronotum?
a. Yes .... Go to #50
b. No .... Go to #52

50 Is the body covered with short, fine hair?
a. Yes .... Allecullidae
b. No .... Go to #51

51 Is the last segment of the antennae about the same size as the other segments?
a. Yes .... Tenebrionidae
b. No .... Lagriidae

52 Are the lateral edges of the pronotum rounded?
a. Yes .... Go to #56
b. No .... Go to #53

53 Is the body long and very flat?
a. Yes .... Cucujidae
b. No .... Go to #54

54 Is the body higher than it is wide, with the head tilted downward to produce a hunch-backed appearance?
a. Yes .... Mordellidae
b. No .... Go to #55

55 Are there two depressions located near the posterior edge of the pronotum?
a. Yes .... Melandryidae
b. No .... Cryptophagidae

56 Are there two depressions located near the posterior edge of the pronotum?
a. Yes .... Melandryidae
b. No .... Go to #57

57 Is the pronotum widest in front and narrower than the elytra in back?
a. Yes .... Oedemeridae
b. No .... Go to #58

58 Are at least the terminal portions of the antennae filiform (threadlike) in shape?
a. Yes .... Go to #59
b. No .... Pyrochroidae

59 Is the pronotum widest at the middle and narrow in the front and back?
a. Yes .... Salpingidae
b. No .... Go to #60

60 Does the abdomen show six ventral segments?
a. Yes .... Meloidae
b. No .... Go to #61

61 Do the hind coxae touch each other?
a. Yes .... Pedilidae
b. No .... Anthicidae

62 Is there a row of conspicuous, flat spines on the outer edge of each tibia?
a. Yes .... Heteroceridae
b. No .... Go to #63

63 Are the antennae distinctly club-shaped?
a. Yes .... Go to #64
b. No .... Go to #69

64 Are the antennae elbowed?
a. Yes .... Scolytidae
b. No .... Go to #65

65 Is the body dark, shiny, oval, and very convex?
a. Yes .... Phalacridae
b. No .... Go to #66

66 Is the head concealed by the pronotum when viewed from above?
a. Yes .... Ciidae
b. No .... Go to #67

67 Are the antennae clavate (segments gradually enlarged toward the tip)?
a. Yes .... Mycetophagidae
b. No .... Go to #68

68 Are the anterior corners of the pronotum curved prominently forward?
a. Yes .... Endomychidae
b. No .... Erotylidae

69 Is the body oval and highly convex?
a. Yes .... Byrrhidae
b. No .... Go to #70

70 Are the antennae longer than half the body length?
a. Yes .... Cerambycidae
b. No .... Go to #71

71 Is the body egg-shaped and widest toward the rear?
a. Yes .... Bruchidae
b. No .... Chrysomelidae

72 Are the elytra short, concealing only about half of the abdomen?
a. Yes .... Pselaphidae
b. No .... Go to #73

73 Is the body shape broadly oval to almost spherical and highly convex dorsally?
a. Yes .... Coccinellidae
b. No .... Endomychidae

74 Does the front of the head extend into a long and narrow or short and broad snout?
a. Yes .... Go to #75
b. No .... Scolytidae

75 Is the snout short and broad?
a. Yes .... Anthribidae
b. No .... Go to #76

76 Is the snout curved downward and are the antenna club-shaped and elbowed?
a. Yes .... Curculionidae
b. No .... Brentidae

Above True to their name, water scavenger beetles, like *Hydrophilus piceus* from England, are mostly aquatic and feed on decaying plants. Unlike predacious diving beetles, with which they may be confused, water scavenger beetles kick their hind legs alternately while swimming and surface head first for air.

BUTTERFLIES & MOTHS
ORDER LEPIDOPTERA

SELECTED FAMILIES

What could enhance the serenity of a sunny meadow full of wildflowers more than the occasional butterfly flitting aimlessly about? A kaleidoscope of colors and patterns, delicate structure, and ethereal flight combine to elevate butterflies above all other insects in the eyes of most people. They are truly the crown jewels of the insect world. Moths, too, enjoy a somewhat heightened status, even though their image is a bit tarnished by the irritating habits of a few species and many are nocturnal and therefore less familiar than butterflies.

Members of the Family Lepidoptera, which means "scale wing" when translated from its Greek roots, are characterized by two pairs of large, membranous wings covered in roof-shingle fashion by tiny scales that account for their often brilliant hues. These colors are the result of either pigments in the scales, or their prismatic structure, or both. The configuration of scales may be such that they refract, or bend, white light to produce blue, iridescent, or metallic colors. Pigments absorb all wavelengths of visible light except for one which is reflected and seen as a given color. Wing scales of butterflies and moths rub off easily, so all specimens should be handled very gently.

At rest, most butterflies hold their wings vertically over the body, while moths hold theirs either in a roof-like slope over the body, or curled around it, or spread flat against the support to which the moth is clinging. All butterflies are active only during daylight, and most moths are nocturnal, although the more colorful fly by day.

The adults of most species have siphoning mouthparts that, at rest, are coiled like a watch spring under the head. These mouthparts can be unfurled into a long tube, or *proboscis*, capable of probing deep into flowers in search of nectar, and certain species also use their proboscis to feed on tree sap or the juices of fermenting fruit. Butterflies and moths are important pollinators of many plants. In some species, the mouthparts of adults are underdeveloped or absent, so these individuals do not eat, and the adults of a few of the more primitive moths use chewing mouthparts to consume pollen. Butterfly antennae are always long, slender, and usually knobbed or hooked at the tip, while those of moths vary but are often feathery or fan-like.

Both moths and butterflies are holometabolous. The wormlike larvae, commonly known as *caterpillars*, have strong chewing mouthparts. The vast majority are plant-eaters, and the habits of a few

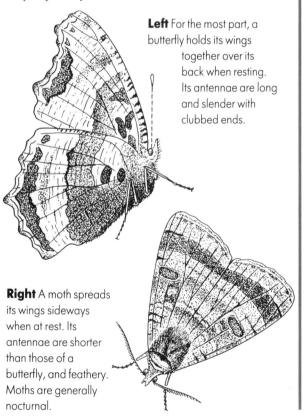

Left For the most part, a butterfly holds its wings together over its back when resting. Its antennae are long and slender with clubbed ends.

Right A moth spreads its wings sideways when at rest. Its antennae are shorter than those of a butterfly, and feathery. Moths are generally nocturnal.

Above With wings held close together over its body, this danaid butterfly (*Danaus eresimus*) from Mexico exhibits a typical butterfly characteristic. Also characteristic are the clubbed tips of the antennae.

Left The wings of this hawk moth (*Batocnema coquereli*), from Madagascar, are held flat across the body exhibiting one of the moth characteristics. Moths are generally drab-colored compared to butterflies. The colors of this rather bright moth break up its outline when on tree trunks.

species have earned them the status of "pest." Caterpillars of most species are restricted to only one species of food plant or to a closely related group of plants.

Caterpillars have a pair of true legs on each of the first three segments immediately behind the head, which together constitute the thorax. Farther back, on the abdominal segments, they also have as many as five pairs of pseudo-legs, or *prolegs*, which are reabsorbed during pupation. Some species are arrayed with stinging hairs or horn-like structures for defense.

The butterfly pupa, which is called a *chrysalis*, is sheathed in a hardened shell that is attached to or near the larvae's food plant. In contrast, the moth pupa is either encased in a cocoon of silk woven by the caterpillar and attached to a support, sheltered in a chamber within a plant stem or underground, or buried beneath leaf litter.

Upon emerging from the chrysalis, most adults set about their primary task, that of reproduction. Lepidopterans are unusual among insects in that males and females of many species exhibit *sexual dimorphism*, meaning a difference in the colors and/or patterns of males and females, indicating that discrimination between the sexes in these species is largely visual. Among those not so endowed, scent disseminated from glands on the wings and/or abdomen is the primary factor in determining sex.

Color also serves as a means of defense for these large-winged insects. Cryptic species often fly at night and are camouflaged against their typical background by day, while those that are brightly colored are advertising the distasteful or toxic nature of their tissues to predators. Many families in this order display "eyespots" on their wings; these really do bear a remarkable resemblance to eyes. It is thought that they also provide a means of defense, because the eyespots may divert a predator's attack toward what appears to be the head but is in fact away from the insect's body. Eyespots can also be flashed suddenly from concealment to startle and frighten small predators.

The span of adulthood in butterflies and moths ranges from about one week to approximately eight months, but the average period for most species is reported to be two to three weeks, during which time courtship, mating, and egg-laying must occur. Courtship can be an elaborate affair, sometimes involving ritualized flights and/or wing stroking between potential mates. Copulation itself is sometimes achieved in flight and may last for several hours.

Eggs, which vary greatly in shape, color, and texture, are normally laid on the preferred food plant of the larvae. In temperate regions, eggs of the last generation of the year either hatch in the autumn, in which case the caterpillars must find a sheltered niche and spend winter in a dormant state in their larval or pupal form, or they hatch during the following spring. A few species over-winter as adults, finding shelter in autumn and entering a dormant period during the colder months.

A small number of species migrate to escape the rigors of winter. One notable example, the monarch butterfly, *Danaus plexippus*, migrates many hundreds of miles to its wintering grounds in southern North America, with the following generations returning to northern breeding grounds the next spring.

Skippers – Family Hesperiidae
Technically neither butterflies nor moths, skippers have some traits of each group. The body of a skipper is thick and bulky, as opposed to that of a true butterfly, which has a long and slender body

---

What you can do

Collecting tips
Capture techniques
aerial net
bait
light trap
Killing method
ethyl acetate killing jar
Preservation techniques
pin adults, wings spread, through
right side of notum

Special notes
1 Do *not* kill members of this family unless you are a serious collector.

Butterflies and moths are one of the few groups of insects whose populations have suffered major declines due to human activities.
2 Reserve a special killing jar for Lepidopterans, so that other specimens do not get contaminated by the lost wing scales.
3 Place each specimen in a triangular paper envelope (see page 55) for protection until they can be pinned.

Above The alert little silver-spotted skipper (*Hesperis comma*) ventures from the woodland edge to a garden zinnia to suck up nectar. Its hooked antennae are characteristic of skippers, and the silver spots on the greenish-tinged underside of the hind wing are typical of the species.

Right The European "scarce swallowtail" (*Iphiclides podalirius*) is not particularly scarce in southern Europe, compared to the ordinary swallowtail, *Papilio machaon*. Its tails and eye-spot mimic the head, compound eye and antennae, thus confusing predators.

in comparison. Skippers are characterized by a wide head and hooked antennae that are widely separated at the base. They can be recognized at a distance by their habit of resting with forewing spread at a 45-degree angle and hind wings spread 180 degrees in a horizontal plane. Wingspans in this family generally range between 14mm and 50mm. Their name can undoubtedly be attributed to their swift, direct, and "skipping" flight.

Swallowtails – Family Papilionidae

Swallowtails are probably the most familiar butterfly family, and possibly the best known family of insects, in the world. They are large, even by butterfly standards, with wingspans of between 54mm and 150mm. Most are either yellow or white, with bold black markings and bright yellow, orange, red, or blue spots, or are black with similar spots. This group is named for the rearward extensions on the hind wings of some species,

which resemble the forked tails of swallows. Parnassians, which lack the tail-like projections of swallowtails, can still be recognized as members of this family by their hind wings, which have only one anal vein.

Milkweed Butterflies – Family Danaidae

Although considered by some to be part of the Family Nymphalidae, milkweed butterflies have antennae completely without scales and therefore, in the opinion of entomologists, differ significantly from them. Milkweed butterflies are fairly large, with wingspans of between 75mm and 102mm. Males have scent pouches on the hind wings and, in some species, brush-like tufts within the abdomen that can be extended to release pheromones that have a tranquilizing effect on females during mating. Most species in this family occur in Asia. North American species are orange or brownish-orange, with black markings; the most famous of

Right The bold orange colors of the North American monarch (*Danaus plexippus*) are a warning to predatory birds that the taste is not to their liking.

Below The clouded yellow (*Colias croceus*) is a very common migratory butterfly. Its pink legs and antennae are a means of camouflaging this butterfly from spiders that lurk in pink-colored flowers, such as thistles.

these is the amazing migratory monarch.

Milkweed butterflies take their name from the larval food plants of North American species. Toxins in the milky sap of these plants are incorporated into the tissues of the insect, imparting a disagreeable taste and causing uncontrollable retching in birds and other predators that ingest them. A number of species outside this family (the Viceroy, for example) mimic the colors and patterns of milkweed butterflies, gaining passive protection for this reason.

Satyrs, Nymphs, & Arctics – Family Satyridae

The wings of satyrs and their kin span between 25mm and 73mm, and are mostly drab gray or brown, with brightly colored eyespots or other markings. Like their close relatives, the brush-footed butterflies, members of this family have front legs that are greatly reduced in size and are not used for walking. The chief distinctions between these two families is that satyrs have from one to three greatly enlarged veins on the front wings. Their erratic, dancing flight takes them

What you can do

Butterfly photography

Butterflies are among nature's most photogenic subjects. It is more difficult to capture these elusive creatures on film than to collect them, but a successful photograph makes a beautiful portrait of the insect in its natural habitat, and it does the butterfly absolutely no harm, an important consideration with less common species. Also, you can photograph all stages of a butterfly's life cycle and its food plants much more easily than you could represent them in a collection.

Successful butterfly photography demands the use of a macro lens, which allows you to focus close enough to obtain frame-filling shots of small subjects. A macro lens should yield at least 1:1 reproduction. That is to say, the image on the film

will be life-size. The major difference between macro lenses is their focal length. A 55mm macro lens can yield about the same image as you will achieve by using a standard 50mm lens to shoot subjects, such as landscapes or people; the difference is that you must be very close to shoot macro subjects. A 105mm macro lens will give you a greater working distance between the camera and the

subject, and a 200mm macro allows even more – no small advantage when working with such wary subjects as butterflies.

Using an electronic flash may help to fill in details and colors, but try to use an exposure as close as possible to that dictated by natural light, so that your background consists of flowers and foliage rather than blackness.

Above This photograph of a *Pieris rapae* in flight shows what can be achieved with a 50mm lens and one extension. High speed films enable you to freeze movement and some of the new automatic cameras will give you a shutter speed of 1/2000 of a second.

weaving close to the ground through the vegetation of their wooded or open, brushy habitats. All caterpillars feed on grasses and/or sedges.

Whites, Sulphurs, & Orange Tips – Family Pieridae

This family runs a close second to the swallowtails in terms of the general public's familiarity with them; in fact, the name "butterfly" was probably derived from the butter-yellow color of familiar European sulphurs. Distribution is worldwide, and they are found over open, sunny areas such as meadows and fields. The names are descriptive of

their typical white, yellow, or orange wing colors, which usually bear either simple black markings or none at all. Orange tips are relatively uncommon and have marbled greenish-yellow, orange tipped wings. With wingspans of between 22mm and 70mm, this is a family of small to medium-sized butterflies. The tips of their otherwise slender antennae are abruptly swollen. Individuals of the same species often exhibit variations in color, and the sexes of many species exhibit sexual dimorphism, a difference in pattern and color between males and females. Some species, such as the Cabbage White, consume food crops.

Gossamer-winged Butterflies – Family Lycaenidae

This family includes those butterflies commonly known as harvesters, coppers, blues, and hairstreaks. Most have dazzling blue, green, violet, or coppery colors, all of which are due to light-bending microscopic structures on the scales. Although similar to metalmarks of the Family Riodinidae, gossamer-winged butterflies hold their wings vertically over the back when at rest, whereas metalmarks usually keep theirs spread flat or at a 45-degree angle. The former are fairly small as butterflies go, with wingspans of from 11mm to 51mm and a brisk, darting flight. Males have front legs that are reduced in size.

Below The colors and patterns of butterflies are fascinating; here, the male of the common blue (*Polyommatus icarus*) shows off its blue iridescence, which is a structural color caused by light refractions.

Right The zigzag patterns, false eye, and two sets of tails on the dusky blue hairstreak (*Calycopis isobeon*), from Mexico, contribute to its head-to-tail mimicry.

Bottom The astoundingly lovely purple-edged copper (*Palaeochrysophanus hippothoe*) from southern Europe takes on a wonderful iridescence in strong sunlight. This specimen was photographed at 6,000 feet in the French Pyrenees.

What you can do

Creating a butterfly garden

Since the middle of the 20th century, the increasing and indiscriminate use of pesticides, combined with accelerated habitat loss, have dealt butterfly populations two heavy blows. By planting a butterfly garden, you can help them while enjoying both their beauty and that of wildflowers. Depending upon your ambition and resources, and the space that you can make available, your butterfly garden can range in size from a few window boxes or a patio border up to a full-blown meadow.

Your first step should be to take a survey of public gardens, parks, meadows, vacant lots, and similar sunny places around your home during spring, summer, and fall and, with the aid of field guides, identify the plants that are already attracting butterflies to the area. These should form the basis of your garden, and to them you can add food plants listed in the butterfly guides for those species normally found in your region that you would like to attract but have not observed. To maximize the effective season of your butterfly garden, select plants with overlapping blooming periods.

You can either buy seeds or plants commercially or gather seeds from wild plants, remembering to observe all local regulations. Seeds gathered in fall should be planted immediately or stored outdoors in a dry place that simulates natural winter conditions as closely as possible. Consult reference books for information on growing native plants from seed. Natural history museums, botanical gardens, and university botany and entomology departments are also good sources of information.

The following spring is the time to plant your garden. Most butterflies prefer sunny, open locations, so keep this in mind when selecting a site. Individual species will seek nectar at different times of the day, so scatter the various plants throughout the garden to ensure that at least some will be sunlit when a given species is feeding. Keeping taller plants toward the perimeter will also help.

In addition to nectar sources for adults, a complete butterfly garden will also include food plants for caterpillars, mud and water puddles, rocky areas, and shelter from rain and wind.

Some popular butterfly food plants

Common Name	Genus	Adult	Caterpillar
alfalfa *	Medicago spp.	X	
apple	Malus spp., Pyrus spp.		X
aspen	Populus spp.		X
aster *	Aster spp.	X	
bee balm	Monarda didyma	X	
beggar ticks	Bidens spp.	X	
blazing star	Liatris spp.	X	
buttonbush	Cephalanthus spp.	X	
butterfly bush *	Buddleia spp.	X	
cherry	Prunus spp.		X
clover *	Trifolium spp.	X	X
columbine	Aquilegia spp.	X	
daisy	Chrysanthemum spp.	X	
dandelion	Taraxacum spp.	X	
dill	Foeniculum vulgare		X
dogbane *	Apocynum spp.	X	
fennel	Anethum graveolens		X
goldenrod	Solidago spp.	X	
grasses	Poa spp.		X
hackberry	Celtis spp.		X
hops	Humulus spp.		X
ironweed	Vernonia spp.	X	X
jewelweed	Impatiens spp.	X	
knapweed	Centaurea spp.	X	
lantana	Lantana spp.		X
lilac	Syringa spp.	X	
mallow	Malva spp.	X	X
milkweed *	Asclepias spp.	X	X
mint *	Mentha spp.	X	
nettle	Urtica spp.		X
parsley	Petroselinum crispum		X
penstemon	Penstemon spp.	X	
plum	Prunus spp.		X
poplar	Populus spp.		X
privet	Ligustrum spp.	X	
rock cress	Arabis spp.		X
self-heal	Prunella vulgaris	X	
sweet pepperbush	Clethra alnifolia	X	
thistle *	Cirsium spp.	X	X
tickseed	Coreopsis grandiflora	X	
vetch	Vicia spp.		X
violet	Viola spp.	X	X
wild carrot	Daucus carota	X	X
willow	Salix spp.		X
winter cress	Barbarea spp.	X	X
wormwood	Artemisia spp.	X	
yarrow	Achillea spp.	X	

* extremely popular with adult butterflies

What you can do

Butterfly watching tools

aerial net
field guides (to butterflies, wildflowers, and trees)
glass or plastic jar
hand lens
blunt forceps
binoculars
field notebook

For the environmentally-minded practical entomologist, butterfly watching is an attractive alternative to collecting.

Note the flight patterns of individuals, as this will often be an important clue to their identity. You may need to capture the butterfly to make a positive identification. With practice, this can be done without harming it. Upon netting a butterfly, give the net handle a quick half twist to trap the insect inside and gently immobilize it as quickly as possible. You may remove your captive from the net by grasping the closed wings with a pair of blunt forceps (never with your bare fingers, as too many scales will be rubbed off) and place it in a jar for observation with a hand lens as you compare it with descriptions and illustrations in the field guide. You will probably want to observe behavior before capturing a specimen, since it may be some time before a released butterfly returns to its normal activities.

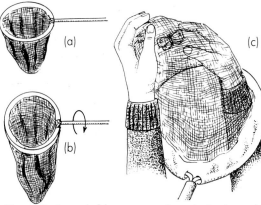

Above At the end of the sweep twist your wrist (a) to fold the bag over the mouth of the net (b). A carrying container can be made by taping a thick piece of black paper over one end of a cardboard tube. Insert this into the net (c), perch the butterfly on your finger and maneuver it carefully into the container. Seal with black cloth and an elastic band.

Metalmarks – Family Riodinidae

Although the majority of this family are colorful species found in the tropics of the western hemisphere, the wings of most temperate region metalmarks are colored in shades of brown, gray, or rust. Many species in the Family Riodinidae also display shiny metallic lines, borders, or other marking. They are known for their habit of perching on the undersides of leaves, with their wings spread flat or at a 45-degree angle. Like gossamer-winged butterflies, males of this family also have vestigial front legs that are not used for walking. Metalmarks are small butterflies, with wingspans of from 16mm to 51mm.

Brush-footed Butterflies – Family Nymphalidae

A very diverse family, brush-footed butterflies have been located in nearly all terrestrial habitats except polar ice caps. They are medium-sized somewhat orange butterflies with wingspans usually ranging between 36mm and 78mm. Most are good fliers, and several migrate or otherwise overwinter as adults. The common name of this family is apparently a reference to their vestigial pair of front legs, which are too short to be used for walking. Brush-footed butterflies are also distinguished by conspicuous knobs on their antennae and by their thick, hairy palpi. The family includes painted ladies, tortoiseshells, admirals, checker-spots, fritillaries, longwings, and crescentspots.

Above Snout butterflies are named after the extension of their mouthparts. There are many species in the Americas, and this one (*Libythea carinenta*) from Mexico is feeding on wild fruit. Typically, snout butterflies hibernate through the winter and emerge in the spring to suck juices from the buds of trees and shrubs.

Right The painted lady (*Cynthia cardui*) is one of the world's most powerful butterflies. During the spring and summer it moves northward in waves from North Africa, through Western Europe. Once it finds a suitable place, the butterfly often defends its own territory. This female is swollen with eggs, which she will lay on thistles.

Above Dusted with iridescent scales, this metalmark (*Caria mantinea*, of the Nemobiidae family), from Peru, displays the typical characteristics of its family. It rests with its wings held flat, rather like a moth.

Sphinx Moths – Family Sphingidae

Sphinx moths, also known as hawk moths, constitute a large and fairly common family. They are stout-bodied and have rigid, narrow, and often boldly marked wings. The front wings are normally about twice as long as the hind pair, spanning between 32mm and 155mm. With extremely rapid wingbeats, these moths resemble tiny hummingbirds as they hover in front of a flower while feeding on nectar. Most are nocturnal or are active at dusk, but some very common species are diurnal. Some have wings with areas devoid of scales and resemble bumblebees in flight.

Giant Silkworm Moths – Family Saturniidae

Much sought-after by collectors, these brightly colored insects are the largest and showiest of moths. Most have wingspans of from 30mm to 150mm, but some tropical species may reach 250mm. Some display very long tail-like extensions on their hind wings. Their antennae are large and feathery, particularly among the males. The adults have only vestigial mouthparts and do not eat. They are short-lived and focus all efforts during their fleeting adulthood on mating and egg-laying. Most are nocturnal and will usually be seen fluttering around windows or outdoor lights. The name of this family comes from their copious use of silk in fashioning cocoons.

Below The elephant hawk-moth (*Deilephila elphenor*) shows off its vermilion colors, enhancing its defensive stance. One of its very large tibial spurs can be seen on its leg, and the fairly uniform antennal segments are clear. Its compound eyes look formidable, but each eye is made up of the usual ommatidia.

Tiger Moths – Family Arctiidae

The bright colors and contrasting black markings on the wings of many species in this family are reminiscent of the large cats after which they are named. Those so colored are usually day-fliers and are advertising the toxic nature of their tissues to prospective predators. At rest, the wings are held in roof-like fashion over their hairy, robust bodies. Fully extended, the wings will span between 12mm and 80mm. Like members of the Family Noctuidae, tiger moths also have a large tympanum on each side of the thorax.

Owlet Moths & Underwings – Family Noctuidae

Noctuidae, the largest family in the order, is composed mostly of drab, medium-sized moths with wingspans usually ranging between 30mm and 50mm. They resemble arrowheads when at rest, with wings folded roof-like over the body. An interesting adaptation of these moths is the presence of prominent paired tympana located on each side of the thorax. The moths themselves make no sound, but these auditory organs are tuned to the high-pitched sounds emitted by foraging bats, forewarning the moth and enabling it to execute evasive maneuvers as a bat closes in.

Above A master of camouflage and disguise, the fully-grown caterpillar of the great peacock moth (*Saturnia pyri*) of southern Europe blends with the leaves that it is eating. It is also armed with blue-topped tubercles which carry defense hairs and spines. The eight pairs of spiracles can be seen along the body, as well as the three pairs of legs.

What you can do

Trolling for moths

Females of many insect species release a type of pheromone, or chemical message, that attracts males, often over great distances. So sensitive are the finely-branched antennae of the males of certain moth species that they can detect pheromone particles in the air at a concentration of only a few parts per million and follow the increasing concentration upwind to its source.

You can lure male moths with a live female as bait in a cage fashioned from fine wire mesh. Use a species in which the males and females exhibit sexual dimorphism – different colors and/or patterns – so that they can be easily distinguished from one another. Leave the caged female outside, preferably in a light breeze, during the normal activity period of that species, and check for males at the end of it.

Females stop secreting their sex attractant after they have mated, so you will need a virgin female for this experiment.

Unfortunately, unless you rear the caterpillars yourself and separate the sexes as soon as the adults emerge, there is no way to tell if your moth is a virgin or not, so you may have to repeat the experiment several times before it succeeds.

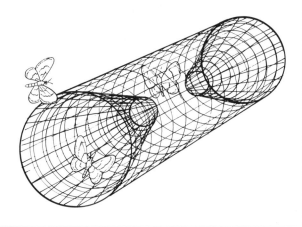

What you can do

Pinning moths & butterflies

Materials
spreading board
insect pins
broad paper strips

Mounting butterflies and moths involves a slightly more complicated procedure than that required for most insects. In order to make a good mount, the wings, critical to the identification of most species, must be properly positioned as they dry. Fully spread, the wings of butterfly and moth specimens will droop over time. To ensure that they end up on a horizontal plane, they are first positioned on a spreading board to dry at a slight upward angle.

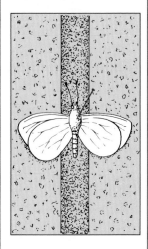

Pin the insect, freshly killed or relaxed, through the right, dorsal side of the thorax and adjust it to the correct height on the pin. Insert the pin in the groove of the board until the base of the wings is level with the top of the board at the groove. Secure the abdomen between two vertical pins to keep it immobile.

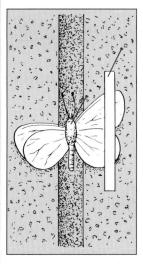

With a pin, gently push against the base of the large vein (costa), moving the front wing forward until its rear edge is perpendicular to the body. Hold it in place by applying gentle pressure to a paper strip placed over the wings, and insert a pin firmly through the paper just ahead of the wing.

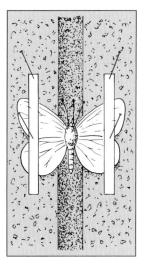

Secure the front wing temporarily with a pin through the paper immediately behind its rear edge. Repeat this procedure for the other front wing.

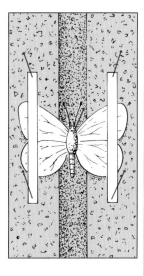

Remove the pin behind the front wing and move the hind wing forward in similar fashion until its front edge just moves under the posterior edge of the front wing. Insert a pin firmly through the paper just behind the hind wing. Repeat this procedure for the other hind wing.

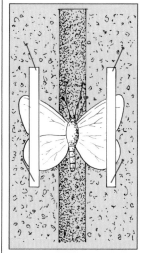

Move the antennae into their natural position and use pins to hold them. If the abdomen sags, cross two pins underneath it to form a supporting cradle.

Drying time will vary with the size of the specimen, temperature and humidity. Larger specimens may take a week or more. Keep specimens in a dry location and out of direct sunlight.

Ctenuchas & Wasp Moths – Family Ctenuchidae

Because of their diurnal habit of visiting flowers, many ctenuchid moths have adopted mimicry of dangerous species as their defense. Some are wasp-like in appearance, with slender bodies, narrow wings spanning some 28–50mm that are at least partly clear (transparent), and rapid wing-beats. Others mimic toxic insects such as tiger moths, or brilliant metallic beetles.

Clearwing Moths – Family Sesiidae

Like the Family Ctenuchidae, the day-flying clear-wing moths also mimic wasps for protection. Most have dark, slender bodies conspicuously marked with red or yellow bands. Their wings, measuring from 13mm to 60mm from tip to tip, are mostly transparent, and fringes of long hair adorn their legs. The sexes of individual species are usually colored differently.

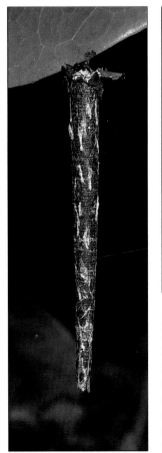

Left The caterpillar of the bagworm moth (Psychidae family), from Penang, Malayasia, lives inside this extraordinary long bag constructed from debris spun in silk. Note the anchoring strand of silk at the side.

Top An insect with a pair of antennae three times the length of its wings, the male long-horned moth (*Adela reaumurella*) has a finely-tuned sense of smell, which it uses for detecting females and aggregating in mixed company.

Above Just a few millimeters long, this is the webworm moth (*Agonopterix arenella*, belonging to the Oecophoridae family), whose wings form a shield shape, slightly overlapping each other.

Gelechiid Moths – Family Gelechiidae

Though small, with wingspans of less than 16mm, gelechiid moths are common, and are fairly easily identified by the often brilliant metallic colors of their wings. The hind wings, which are wider than the front wings, are pointed, and have a broad hairy fringe along the outer edges.

Geometer Moths – Family Geometridae

The larval forms of these moths are the famous "inchworms" or "measuringworms" that travel with a looping motion of their bodies, pulling the rear ends up to the thoracic legs, grasping their support with posterior prolegs, and extending their bodies forward. Adults, though common, are rather non-descript with slender bodies and broad, delicate wings spanning about 8–65mm. They are most easily distinguished from similar families by their habit of resting with their wings fully spread, rather than folded.

What you can do

Insect condos

Galls are swollen masses of plant tissue caused by another organism. Examine stands of goldenrod in weedy areas during late summer. Very common on these plants are galls on the stems, where insects have laid eggs and the chemical secretions of the developing larvae have stimulated a swelling of the plant tissues around the site. If there is no hole to show where the adult has emerged, then the larvae or pupae are still inside. You can cut away galls and keep them in separate jars at home to see what emerges. Elongated, spindle-shaped galls are likely to yield gelechiid moths, while round galls will probably produce a fly from the Family Tephritidae.

Above The wavy marks on the wings of the codling moth (*Cydia pomonella*) are interrupted at the tips by a semi-metallic dusting in the shape of a thumb mark. The larvae are major pests of apples.

Above The brown house moth (*Hoffmannophila pseudospretella*, of the Oecophoridae family) is a common insect, about 10mm long, which lays its eggs in the dust and debris found in houses. Its wings are covered with silvery-gray scales.

FLIES
ORDER DIPTERA

SELECTED FAMILIES

Nobody likes them. Some are parasitic; others bite, contribute to the spread of diseases, or are simply downright annoying. So what can we say about flies that could possibly be interesting? For one thing, they are important prey for a great many animals, and for this reason alone we must tolerate them. They are also invaluable as pollinators, second only to bees and wasps, and are effective scavengers and nutrient recyclers. Since we have to live with flies, we might as well learn more about them.

Unlike the vast majority of other species, flies, of the Order Diptera, which means "two wings," have only one pair of wings. The single pair of membranous wings are attached to the mesothorax, which, enlarged and packed with flight muscles, is responsible for the high speed and wingbeat frequency of many fly species. In contrast, the prothorax and metathorax are much smaller. A fly's hind pair of wings, while not completely absent, are reduced to tiny vestigial appendages called *halteres*, which extend from the metathorax. Many families of flies have a membranous lobe at the base of each wing that covers the halteres. The presence or absence of this lobe, called a *calypter*, is often a clue to the family.

Halteres are thought to function as stabilizers in flight. Sensory organs at their bases read air currents, and the information they relay to the brain gives the fly amazing control and maneuverability, far greater than that of any other family of insects. These, combined with the claws and pads on their feet, enable them to land on virtually any surface, even ceilings!

Most adult flies have sucking mouthparts that are adapted for either sucking, sponging, or lapping liquids, although in a few species they are reduced or absent. The majority of species feed on nectar, but a significant number consume blood, plant sap, fruit juice, other insects, or decaying organic matter. Nectar feeding flies generally visit many different species of flower, but the nectar in flowers with deep corollas is inaccessible to them.

Flies engage in complete metamorphosis in the course of their life cycle. Eggs are laid individually on or near the larval food source. This is often soft, moist, decaying material, although the larvae of certain flies consume significant numbers of agricultural pests. In most species, the larvae are soft-bodied, with no legs and an indistinct head, and are commonly referred to as *maggots*. Those that are aquatic – mosquitoes and midges, for example – are much more mobile and have well-defined heads. Among the higher flies, the pupae are encased in their last larval skin, which has become a tough, impermeable capsule called a *puparium*, resistant to environmental extremes. Pupae often require precise conditions before they can develop into adults.

What you can do

Capture techniques
aerial net
sweep net
aspirator
bait trap
malaise trap (p.131)
Killing method
ethyl acetate killing jar
Preservation techniques
pin through right side of notum
point (mount insect on its side)

Note When sweeping vegetation, slow-flying specimens can be caught with short, low strokes of the net, passing just ahead of your feet; wide strokes, passing farther ahead and just over the tops of weeds and grasses, will net some of the faster species.

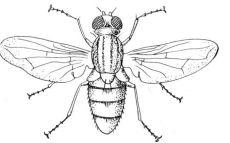

Above Midges belonging to the Chironomidae family swarm around leaves at dusk, aggregating by virtue of their pheromones, and cavorting in various prolonged dance displays. Courtship and mating eventually ensue. Their fast and furious wing-beat produces a high-pitched whine which is also probably used for aggregation purposes. Midge larvae breed mostly in freshwater.

Above Unlike most other insects flies only have one pair of wings. The reduced hind pair give the fly more maneuverability and control than any other family of insect; by acting as stabilizers that accurately read air currents.

Crane Flies – Family Tipulidae

Looking for all the world like huge mutant mosquitoes, crane flies cannot deliver the formidable bite of which they appear capable. Aside from their extremely long, stilt-like legs, they are easily distinguished from most other flies by their large size (8–65mm in length), slender body, and long and narrow wings, which protrude widely from the body when at rest. Adult crane flies are especially common in moist areas of dense vegetation, although they frequently enter buildings.

Mosquitoes – Family Culicidae

That high-pitched hum in a darkened bedroom is irritating in itself, and the anticipation of a bite dashes any hope of sleep! Most mosquitoes are less than 6mm long, but few insects are involved more directly in an adversarial relationship with humans. True, they are the vectors of some of our more serious diseases, but in temperate regions they are, for the most part, merely bothersome. Much has been made of the fact that only female mosquitoes bite, while males feed on plant juices and nectar and are important pollinators of many wildflowers. Females also consume nectar, but must have blood in order to obtain the protein necessary to nourish their developing eggs. Males are most easily recognized by their relatively feathery antennae.

Mosquitoes lay eggs in water, and many species can reproduce in almost anything that will hold water for the week or so it takes them to develop. Some actually show a preference for breeding in man-made containers. The larvae, called wrigglers, are quite active, and have a distinct head but no legs. They hang upside down at a 45-degree angle or float horizontally just under the surface and breathe through an abdominal tube that pierces the surface film. Pupae, called tumblers, are less mobile but still active swimmers, rising to the surface to breathe through two small tubes on the notum.

Horse & Deer Flies – Family Tabanidae

At least mosquitoes, with their high-pitched hum, give warning as they approach. Such is not the case with horse flies and deer flies, which fly swiftly but quietly, landing softly before inflicting a

Left The horsefly's amazing compound eyes, here seen in close-up, are remarkable for their iridescent combination of green and purple. The rows of minute ommaditia are clearly visible. The females feed on blood and then only to supply their developing eggs with protein.

Above The aquatic larva of the mosquito *Anopheles atroparvus* has several pairs of gills, through which it extracts oxygen from the water. Its body is divided clearly into head, thorax and abdomen, demonstrating these segments well. The larva defends itself by contracting and relaxing its muscles and thus wiggling away from danger.

Left Mating craneflies (*Tipula varipennis*) are a common sight in the autumn, and this pair display the main feature that makes them true flies – the halteres, or modified hind wings, read wind currents and function as stabilizers.

painful bite. Here again, it is only the females who bite, and only when they need protein for their developing eggs, while males subsist solely on nectar and pollen.

These are robust, wide-headed flies, 6–28mm long, with huge eyes that are often brilliantly colored or iridescent. The eyes of males touch one another, while those of females are separate. Horse flies are the larger of the two, and have a prominent spur on their third antennal segment. Deer flies usually have darkly-banded wings and no spurs on their antennae.

Robber Flies – Family Asilidae

Robber flies are predatory and often prey upon resting insects larger than themselves, attacking from above and using short but stout mouthparts to pierce the tough exoskeleton of their victim and drain its body fluids. Some employ a wolf-in-sheep's-clothing approach by mimicking bumble-bees, which are non-predatory. Such an appearance undoubtedly protects the robber fly as well as facilitating its advance on its prey. A few others mimic damselflies for the same advantage.

Robber flies are marked by a long, spindly

Above A male midge has a pair of large brush-like antennae which make it well equipped for detecting odors.

Right Showing its piercing mouthparts this large member of the Empididae family (*Empis tessellata*) is feeding on the flowers of buckthorn. However, most of the time they catch and kill prey by sinking their mouthparts into the bodies of their victims.

What you can do

Build a better fly trap

When they encounter an obstacle in flight, many flies will change direction and fly *upward*. You can use this item of knowledge to your advantage and construct fly traps that will collect many different specimens for you.

A classic tool of entomologists is the malaise trap, a house-shaped affair, about two yards high and the same or more long, made from the same soft mesh that you would use to make an aerial net. A large square or rectangle forms the roof, and a five-sided piece forms the side on each of the "gabled" ends. Instead of side walls, this house has one sheet of netting, three sides of which are sewn to the peak of the roof with each end hanging down the middle to form a baffle that flies will encounter. Use lead fishing sinkers at the bottom of the mesh baffle and ends to keep them vertical in a breeze. Slope the top of the roof somewhat, leaving a hole at one end into which a small jar or vial can be inserted upside down and sealed with a rubber band. It is here, at the highest point, that your captives will tend to congregate. Your malaise trap can be supported by

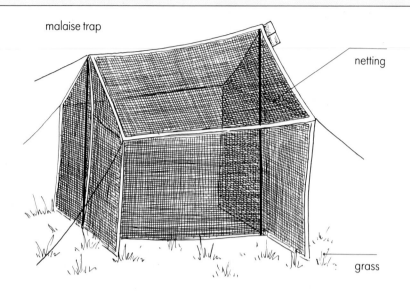

malaise trap

netting

grass

either a pole at each end or a line tied between two trees, with guy lines staked from the roof corners and peaks to hold it open.

A simpler but effective trap can be made by placing a small amount of bait in a jar lid or other such container set on the ground. Directly above this place a large white plastic funnel, upside down and supported at three points by sticks, stones, or whatever, so that it is half an inch or so above the ground. Top this with a tall, clear glass jar placed upside down over the funnel. The jar mouth must be narrower than the widest part of the funnel.

Yet another type of fly trap can be made from a half-gallon plastic milk jug. A little less than an inch above the bottom, cut several holes about half

simple funnel trap

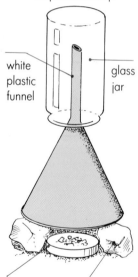

white plastic funnel

glass jar

jar lid with bait   supports

an inch in diameter. In the lid of a small screw-top glass jar, cut a hole to match the opening of the jug; seal this lid (top side down) on top of the jug with silicone bathtub caulk. Place a small amount of bait in the jug;

screw the jar onto the lid, and suspend the trap by the jug handle, hanging it from a tree branch or other support, so that it is at roughly a 45-degree angle, with the jar at the top.

The last two traps can be set with various baits. Carrion, dung, or overripe fruit will attract certain families. The recipe below for a fermented sugar bait will be attractive to moths and butterflies as well as flies.

Ingredients

1 cup low-alcohol beer
½ cup brown sugar
1 tablespoon rum
1 tablespoon honey
¼ apple finely diced and dried

Mix ingredients and let stand for 12–24 hours. Multiply ingredients 2–5 times for larger batches.

abdomen that tapers rearward, spiny legs, a bristly, bearded face, and a conspicuous concave depression on the top of the head. As flies go, they are rather on the large side, ranging between 5mm and 30mm in length.

Hover Flies – Family Syrphidae

These are also known as flower flies, both names being equally descriptive. Very common flies, they have a swift, darting flight, interspersed with frequent periods of hovering. They are quite often seen hovering over or resting on flowers, where they feed on nectar. Many strongly resemble bees or wasps in markings and behavior, but they have the distinctive head of a fly, with very large eyes that seem to occupy the entire head. Hover flies do not sting or bite, but run an effective bluff. At rest,

there is a telltale nervous quivering of the abdomen. The larvae of many hover fly species are voracious predators of aphids.

Vinegar Flies – Family Drosophilidae

These are the infamous "fruit flies" that are employed in classic genetics experiments in heredity. In actuality, true fruit flies are members of another family, Tephritidae, but the name is equally appropriate to members of the Drosophilidae, because they frequent the vicinity of ripe or fermenting fruit and decaying vegetation. The larvae live in this type of medium, feeding on the yeasts found there. Adults are stocky but tiny – from 2mm to 4mm long – and a yellowish-brown in color. They, too, feed on yeasts on the surface of fruits and flowers.

Right The drone fly (*Eristalis tenax*) is a powerful fly that is known to migrate. As its name suggests, it mimics the drone honeybee in many morphological ways, including having a fairly large and shiny thorax. It belongs to the Syrphidae family, which is the largest of all fly families, and spends much of its time basking on flowers and leaves. Its larvae are carnivorous, eating aphids and other insects. Syrphids are also called hover-flies since they often hover over flowers before landing.

What you can do

See them smell

Antennae are but one of the means by which insects smell. Many have chemically sensitive *setae*, or hairs, which serve the same function. On many flies and other insects, such setae are concentrated on the tarsi, or feet.

To observe the sensitivity of these chemical sensors, mix solutions of sugar-water in concentrations of 0.5 to 10 per cent in 0.5 per cent increments (10g of sugar in 90ml of water equals 10 per cent, and so on). Capture a housefly, blow fly, flesh fly, or tachinid fly,

then keep it captive for a day or two to ensure that it builds an appetite. Stick a 1cm square of double-backed adhesive tape on the end of a small wooden dowel rod or a new, unsharpened pencil, and gently touch this to the back of the fly. Using the dowel as a handle, hold the fly about 5–6mm from the sugar solutions, starting with the weakest concentration and progressing to stronger solutions. When you reach the first solution detectable to the fly's chemical receptors, its mouthparts will dart downward in an attempt to reach the food.

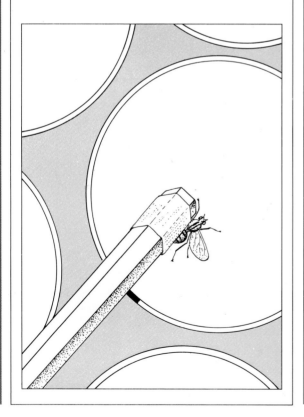

Top The March fly or St. Mark's fly (*Bibio marci*) can be very numerous in meadows and wet fields in the spring. Their larvae have primitive insect characteristics, and they feed on plant roots and decaying vegetable matter. The adults are recognizable because of their black color and the casual way in which they fly or get blown about the fields.

Above Stratiomyid flies (those belonging to the Stratiomyidae family) often have white, yellow or green markings. In this species of soldier fly (*Stratiomys potamida*) the markings are a delicate lemon yellow. Eggs are laid on vegetation close to water, and the developing larvae eat decaying vegetable matter as well as other small insects. The adults are poor fliers and frequently congregate on the umbels of umbelliferous plants.

House Flies & Kin – Family Muscidae

Possibly the best-known insect species in the world is the common housefly, *Musca domestica*, a member of the Family Muscidae. Other members of this family are much like the housefly in appearance and size (3–12mm long), and they differ from similar families, such as tachinid flies, flesh flies, and blow flies, which have stout bristles on the sides of the thorax above and below the wing bases, a feature lacking in muscid flies.

Many in the Family Muscidae feed with sponging mouthparts. A large, fleshy mass at the tip of the proboscis absorbs liquids and food particles, which are then siphoned through the proboscis. They can spread disease-causing bacteria by walking on human food, utensils, or other surfaces after having visited unsavory materials, such as manure. A few muscids have piercing-sucking mouthparts and can bite.

Blow Flies – Family Calliphoridae

Roughly the size of houseflies – about 5–15mm long – blow flies are very often metallic blue or green in color. Close inspection with a hand lens will reveal a feathery projection, the *arista*, on each of the short antennae. These are quite common flies, many of which both feed and lay their eggs on carrion or dung. A few species lay eggs in open sores, wounds, or even the nostrils

of live animals (hence the name), and the resulting maggot infestation can cause the host severe irritation and weakness.

Flesh Flies – Family Sarcophagidae

Flesh flies are often identifiable by the combination of alternating black and gray longitudinal stripes on the thorax, which are absent among other flies, and a checkered abdomen. They are never metallic. Their body length ranges between 2mm to 14mm. Many lay their eggs in carrion, dung, or open wounds, although the larvae of some species are internal parasites of other insects, particularly grasshoppers and beetles.

Tachinid Flies – Family Tachinidae

Tachinid flies differ from other similar families in having a large postscutellum, a prominent lobe that occurs on each side of the thorax below the scutellum, which is the triangular sclerite behind the mesonotum. Their bodies are 3–14mm long, usually stocky, and often densely studded with stout bristles. The larvae of this large family are internal parasites of many other insects, and are second only to parasitic wasps as nature's insect population control. The eggs are either deposited directly on the host or on plant tissue which is then eaten by the host. Tachinid fly larvae usually kill their host just as they themselves become ready to emerge and pupate. Adults feed on nectar.

Above The bee fly (*Bomylius major*) is easy to identify since it has a long proboscis which it uses to probe flowers for nectar. This fly makes a high pitched whining sound, similar to that of a bee.

Above Dung flies (*Scopeuma stercorarium*) swarm on fresh dung and often mate while at least one is still feeding. Eggs are deposited on the dung, and the larvae develop rapidly.

Above Flesh flies (Sarcophagidae family) can be recognized by the strong black or gray longitudinal markings that pass down the thorax. Females incubate the eggs in their abdomens and lay live larvae. The larvae live on decaying animal or plant matter.

ANTS, BEES, WASPS, & KIN
ORDER HYMENOPTERA

SELECTED FAMILIES

The ecological significance of hymenopterans cannot be overstated. Parasitic wasps are the primary agents of insect population control. Ants aerate and fertilize the soil, recycling as many soil nutrients as do earthworms. Bees are the major pollinators of flowering plants, which in turn feed the world's terrestrial animals and provide the oxygen we breathe. In fact, about 150 crops throughout the world are pollinated mostly or entirely by bees and, as a bonus, they provide us with honey, a major crop in itself.

Entomologists consider this order to be the most advanced among insects, due to their highly specialized nature. Ants and some families of wasps and bees have evolved cooperative social hierarchies in which there is a division of labor among different castes. In such situations, workers are sterile females who care for the eggs, larvae, and pupae, maintain the nest, and repel invaders. The few males of the colony exist only to mate with the queen, the only member of the colony to lay eggs, which is her sole function.

Even though the best known hymenopterans are colonial, most species are solitary. Many of these build nests from mud or excavate nests underground or in wood or plant stems. Solitary wasps usually provision their nests with freshly killed or paralyzed insects or spiders, while non-colonial bees stock theirs with pollen and nectar. It is thought that, even though the parents and offspring never see one another, provisioning by the parents is the precursor to true social behavior, in which the adults of one generation cooperatively feed and care for immature insects of the next.

Some other solitary species do not build nests, but lay their eggs in or on the eggs, larvae, or pupae of other insects. These are generally termed parasitic, although they actually kill their only host. They can more correctly be called *parasitoids*, for they fall somewhere between true parasites, which do not kill their host, and predators, which kill more than one victim. If the immature wasp feeds from the outside of its host, the adult female will first sting and paralyze the victim before depositing her egg.

bee

ant

Above and **left** Both bees and ants have two pairs of transparent wings (with the exception of the wingless worker ants), "wasp waists" and chewing mouthparts. The best known species are social and live in colonies but the majority are solitary.

Probably the main factor in the evolution of most hymenopteran lifestyles was the development of the "wasp waist." In all but the more primitive families, namely sawflies and horntails, the first abdominal segment, or *propodeum*, is separated from the rest of the abdomen by a highly flexible and often constricted hinged joint, the *pedicel*, which allows precise movement of the abdomen in egg-laying and defense, and also permits the insects to turn around in the tight confines of a burrow or nest.

Other features that distinguish members of the Order Hymenoptera are their chewing mouthparts, 5-segmented tarsi, and four membranous wings (when present). The forewings are somewhat larger than the hind wings, and both pairs generally have few veins and large cells, the spaces bordered by veins. Many have chewing-lapping mouthparts, in which the labium and maxillae are modified into a tongue-like apparatus for drinking liquids. All species undergo complete metamorphosis. Among those bees, wasps, and ants that sting, the stinger is actually a modified ovipositor, or egg-laying structure. Sawflies and horntails, despite their ominous appearance, do not sting.

What you can do

Capture techniques	Preservation techniques
aerial net	pin through right side of
sweep net	notum point
aspirator	70 per cent alcohol vial
bait	

Killing method	Note Many
ethyl acetate killing jar	hymenopterans can sting,
immersion in 70 per cent	and severe allergic
alcohol	reactions may occur in
	sensitized individuals.
	Handle with care!

Left It is comparatively easy to determine whether ants, bees and wasps belong to the three basic groups of hymenopterans. However, it is much more difficult to place the solitary bees and solitary wasps within their individual groups. In addition, there are true flies that mimic bees and wasps, further confusing the matter. This is the wall mason wasp (*Ancistrocerus parietinus*), which is a British solitary wasp.

Sawflies – Family Tenthredinidae

Tenthredinidae, the largest family of sawflies, includes most of the species commonly encountered. Black or brown, and between 3mm and 20mm long, they are most easily distinguished from other sawflies by their thread-like antennae, which have from 7 to 10, but usually 9, segments. As with all other sawflies and horntails, the junction of abdomen and thorax is broad, unlike the more familiar families of hymenopterans. Females have a saw-like ovipositor with which they insert eggs into plant tissues, and they are often colored differently from males of the same species.

Females of many sawfly species, as they lay an egg, inject a substance into the plant that stimulates gall formation, a proliferation of plant cells in the tissues around the egg. This serves as both shelter and as a protein-rich food source for the larva.

Horntails – Family Siricidae

The absence of a "waist" gives horntails a cylindrical appearance, but otherwise they are rather wasp-like. Their bodies are usually from 20mm to 40mm long, brown or black, and often marked with yellow. The name stems from a long, spiny projection on the last abdominal segment of both males and females. Beneath this, females are equipped with a long, stout ovipositor.

Ichneumon Wasps – Family Ichneumonidae

Ichneumons constitute a large family of parasitic wasps embracing a great variety of sizes and colors. Most are uniformly colored, from black to yellow, but some are boldly patterned or have conspicuous white or yellow segments in the middle of their long antennae. The antennae have more than 15 segments and are at least half as long as the body. These are slender insects, often with abdomens that are quite narrow at the base and expand steadily rearward. Many females trail long, thread-like ovipositors behind them, some of which have taste receptors to help them identify suitable hosts. Ichneumons resemble the smaller, stockier parasitic wasps of the Family Braconidae. The larvae of both families are major parasites of other juvenile insects, especially those in the Order Lepidoptera, although they also attack larvae of many beetles and of their own order.

Above Sawfly larvae are gregarious, feed voraciously and, in the case of this gooseberry sawfly (*Pteronidea ribesii*) are commercial pests.

Above With her long ovipositor, (seen here between her second and third pair of legs) buried deep in the wood, a female of the greater horntail (*Urocerus gigas*) furnishes each hole with an egg.

Above Using a combination of vibrations in the wood and her antennae to sense them, the female of the parasitic wasp, *Rhysella approximator*, positions her ovipositor in the correct place to pierce the larvae of the woodwasp (*Xiphydria camelus*) and lay her eggs in them.

Left The ichneumon wasp (*Pimpla instigator*) is principally a parasite of butterflies. The wasp lays its eggs through the cuticle of the caterpillar, and the wasp's larvae feed on the non-vital parts of the body without killing their host. When fully grown, they simply eat their way out of their host and pupate on the writhing remains.

Velvet Ants – Family Mutillidae

Despite their name, these are not ants at all, but very hairy parasitic wasps. They differ from true ants in that they are covered with dense, brightly colored hair, their antennae are not elbowed, and the pedicel at the base of the abdomen is broader than that of an ant and lacks a dorsal hump. Most have blazing red, yellow, or orange patterns and are from 6mm to 25mm long. Females are wingless, smaller than males, and can deliver a very painful sting. Their larvae are external parasites, mostly upon the larvae and pupae of bees and other wasps.

Ants – Family Formicidae

"The ants go marching one by one. Hurrah! Hurrah!..." begins the children's song, reflecting the air of order and cooperation that is synonymous with ants. All ants are social insects, living in colonies with well-defined castes and divisions of labor. Most colonies are composed of a large worker caste of wingless, sterile females and a reproductive caste of winged, fertile males and females, the latter also known as queens.

Workers are aptly-named, for they perform every function critical to the survival of the colony except for reproduction. Building and maintaining the nest, repelling intruders, gathering food, and tending eggs, larvae, pupae, and the queen are all tasks which fall to the workers. Most ant species are scavengers, but many will harvest leaves and seeds, raise fungi in subterranean chambers, prey upon other invertebrates, or herd and protect aphids while harvesting their honeydew secretions.

An ant nest is a maze of tunnels underground or in decaying wood. Workers make regular forays in search of food, and in order to find their way to a recently discovered food source or back to the colony, they lay down a scent trail as they go. In addition, each colony has its own distinct odor, which largely determines recognition among members and rejection of outsiders. In spring and fall, some members of the reproductive caste leave the colony in a short mating flight. After mating, males die and females lose their wings and usually begin a new colony, each caring for the first brood of workers by herself.

Ants are usually black, brown, or reddish, and they range between 1mm and 25mm in length. Their distinguishing features are unmistakably elbowed antennae and a slender pedicel between the thorax and abdomen that, when viewed from the side, reveals one or two distinct humps. Many species are able to spray formic acid or other noxious chemicals as a means of defense.

Vespid Wasps – Family Vespidae

Vespid wasps include the familiar yellow jackets, hornets, and paper wasps, among others. They are from 10mm to 30mm long and have conspicuously notched eyes. Most are brown or black, and some species have distinct white or yellow markings. Their wings have a pleated appearance when folded over the back at rest, a trait that distinguishes them from the similar sphecid and spider wasps. Many vespids are colonial and lay eggs in combs of hexagonal cells made of a papery substance. Such nests may be open or enclosed

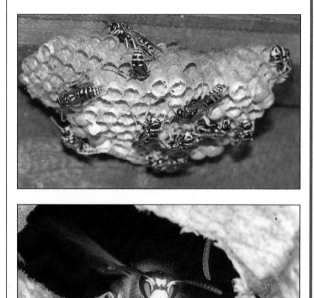

Top Nests of the *Polistes* wasp, from southern Europe, are never very large, are constructed out of wood fiber and are found on buildings, in stone walls or, rarely, in vegetation.

Above A freshly-emerged specimen of the common wasp (*Vespula vulgaris*), complete with all its body hair, looks out of the hive and defies any robbers which might fancy mounting an invasion.

with only one entrance near the bottom. Other vespids are solitary, and lay eggs in underground cells or cells constructed of mud. The cells are provisioned with chewed insect prey rather than whole insects. Yellow jackets and hornets employ the same sort of caste system as do ants and bees, each colony having only one fertile female and a large caste of infertile female workers, whereas colonies of paper wasps consist of males and a group of fertile females living cooperatively in the same nest. All species in this family can inflict painful stings, and alarm pheromones released by disturbed individuals can incite gang attacks.

Spider Wasps – Family Pompilidae

These solitary wasps are typically seen running across the ground while nervously flicking their dark or amber-colored wings. Many are a shiny bluish-black color or have red or yellow markings. They measure from 10mm to 50mm in length and have relatively long legs, with the hind femurs extending to the tip of the abdomen. Spider wasps take their name from the females' habit of provisioning their underground nests with paralyzed spiders, on which the eggs are laid. Despite this habit, the adults usually feed on nectar.

Sphecid Wasps – Family Sphecidae

Sphecid wasps are solitary, and are characterized by a short, collar-like pronotum, with a lobe projecting backward on each side. They are from 10mm to 55mm in length and, unlike those of the similar spider and vespid wasps, their wings do not fold over the abdomen when at rest. The nests, provisioned by females with insects or spiders, are usually either mud cells attached to buildings or

Left A mud-daubing wasp (*Sceliphron formosum*) collects mud from a tiny puddle ready to make her nest of mud on a ceiling or overhang. Small rows of mud nests are made together, and each is furnished with pollen and an egg.

Above The ingenious spider hunting wasp (*Anoplius infuscatus*) outwits the spider and paralyzes it, prior to taking it away to furnish its nest, where it will place an egg on the living food supply. The wasp is a member of the Pompilidae family.

Ant farming

Materials for alternative design

2 pieces glass or plexiglass, 12in × 13in (30.5 × 33cm)
4 pieces wood, 12in × ½in × ½in (30.5 × 1.5 × 1.5cm)
1 piece wood, 4in × 14in × ½in (10 × 35.5 × 1.5cm)
small wood screws
2 large rubber bands
funnel
camelhair artist's brush
cotton
fine sandpaper
several vials
garden trowel
large brown paper envelope

Granted, keeping an ant farm does have a certain childish connotation, but it's really not just an occupation for children. You can maintain a thriving ant colony inside your own home and observe them as they go about their daily lives in the confines of a self-contained system.

One of the simplest ant farms to build requires two straight-walled glass containers, one slightly smaller than the other so that it nests inside the larger vessel with a gap of about ½in between the two around the entire circumference. The inner jar should be slightly shorter, so that the outer one can be capped.

1 Use a bit of modeling clay or similar adhesive substance to hold the inner jar precisely in place while you carefully fill the gap between the two with loose, very sandy soil that has been sifted to remove stones and other debris.

2 Bring the soil level to just below the top of the inner jar and tamp it down gently with the eraser end of a pencil.

3 Food and a piece of damp sponge for moisture can be added to the inner jar. After introducing ants to the formicarium, cover the mouth of the outer jar with fine nylon mesh held in place by a rubber band.

Alternative design

1 Another successful design requires that you first construct a U-shaped frame from three 12in lengths of wood that are each ½in square in cross section.

2 Place the bottom ends of the two pieces forming the sides of the U on *top* of the bottom piece, *not* aside of it.

3 Sandwich this between the two sheets of glass or clear plexiglass with the excess extending past the open end of the U. Any sharp edges should be sanded first. Plexiglass can be drilled and attached with wood screws, but glass is best secured with two heavy rubber bands.

4 Drill a ½in hole in each side of the U, centered about 1½in below the top of the wood, and plug these holes with wads of cotton.

5 Mount the frame, open end up, in the center of a 4 × 14in wooden base, for stability.

6 Fill with soil as described above, tamping it down periodically, until it is about level with the bottom of the holes.

7 Finally, attach a fourth length of wood, with the same dimensions as the other three sides, between the glass or plexiglass on top, and drill a ½in hole in the middle. This design lends itself well to expansion, as you can link two or more ant farms by running clear plastic tubing, available from aquarium supply stores, from one side hole to another.

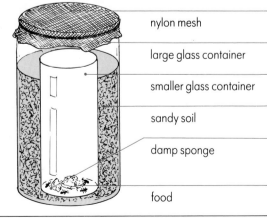

nylon mesh

large glass container

smaller glass container

sandy soil

damp sponge

food

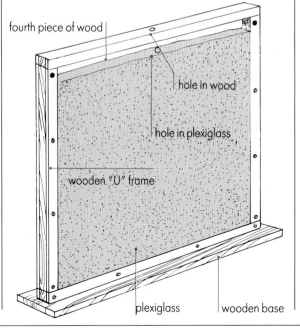

fourth piece of wood

hole in wood

hole in plexiglass

wooden "U" frame

plexiglass | wooden base

Finding the ants

Try to get your ants and soil from the same vicinity. Look for ant colonies under large, flat rocks in vacant lots, meadows, or other areas not recently disturbed. Bait several vials with honey and lay them open, near to the colony. When you have about a hundred ants in the vials, plug the openings with wads of cotton. Next, unearth the colony with a garden trowel, placing each scoop of soil on a large sheet of white paper and sifting through it until you find the queen, who will be much larger than the rest. Put her in a separate vial.

Putting ants into the farm

First place a dinner plate upside down on a large tray. Pour enough water in the tray to form an island, from which the ants cannot escape. Place the ant farm on the island and put a small funnel in the top hole. Coax the queen from her vial and down the funnel with a camel hair artist's brush. Introduce the rest of the ants in similar fashion, and plug the hole with a wad of cotton. Those that climb out of the funnel can be trapped again by the same method as before.

The cotton stoppers will allow a free flow of air into the ant farm. Your ants will also need water, which can be supplied with a bit of damp sponge every few days. Since most ants are scavengers, you should have no trouble finding foods they like. Bread and cracker crumbs, bits of fruit or ground meat, and honey will usually be taken readily, but be careful not to overfeed. Also, since ants prefer darkness in the nest, cover the ant farm with a large, dark envelope or other material to keep them content, and try to use artificial light while observing them.

To retrieve pieces of sponge and unwanted food with greater ease, you may wish to establish the feeding area in a screw-top glass jar, connected to the farm with a piece of clear plastic tubing. Once the nest is established and the queen has begun laying eggs, the ants will consider it home and return to it freely. You can run a plastic tube from the farm and out of a window to the ground, allowing them to come and go as they wish, gathering their own food outdoors.

Your ant farm will lend itself to many interesting experiments. Observe the different reactions when you remove a few ants from the colony and later return them, compared to the colony's response when

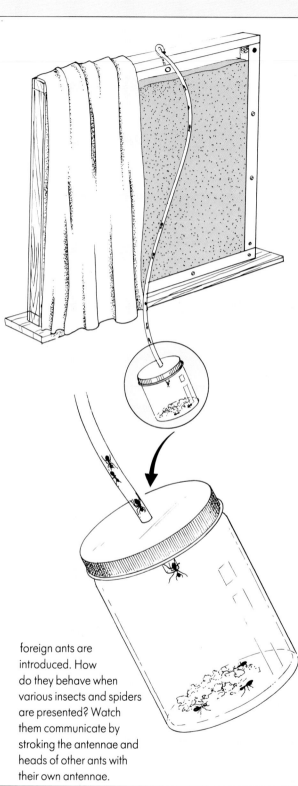

foreign ants are introduced. How do they behave when various insects and spiders are presented? Watch them communicate by stroking the antennae and heads of other ants with their own antennae.

rocks, or are found underground or in natural cavities. Some sphecid wasps exhibit an elementary form of social behavior, in which small groups of females cooperate in constructing a mud nest, although each provisions her own cells. This behavior is thought to be the immediate precursor to true insect social organizations.

Bees – Superfamily Apoidea

Because of their many similarities, all families of bees may be conveniently considered in one superfamily, which consists of the Families Colletidae, Halictidae, Andrenidae, Melittidae, Megachilidae, Anthophoridae, and Apidae. These are separated primarily on the basis of differences in mouthparts and wing vein patterns. By far the most significant is Apidae, which includes bumblebees, carpenter bees, and honeybees. Honeybees and bumblebees are the only social

bees. Others are solitary, and most build nesting tunnels either underground or in wood or plant stems, provisioning them with nectar and pollen balls, on which the eggs are laid. Bees range from 4mm to 25mm in length and are often black or brown, with white, yellow, gold, or orange bands.

Bees are highly specialized for gathering nectar and pollen. The long proboscis is ideally adapted for probing deep into flowers for nectar, and pollen sticks to their hairy bodies as they brush against the anthers of flowers. Pollen brushes and combs – stiff hairs located on the hind legs of females – are used to comb pollen from the body hair and transfer it to pollen baskets, located on the hind tibiae. Pollen baskets sometimes become so overloaded that the bees have difficulty flying. Coincidentally, bees are the primary pollinators in the world, and many plants are completely dependent upon them for this service.

Left The leaf-cutter bee (*Megachile willoughbiella*) methodically cuts out a precise section of leaf, flying off with this held between the legs. The bee makes a nest-cell from the curled piece of cut leaf, which is then furnished with a ball of pollen on which the eggs are laid. In structure and color the leaf-cutter bee is very similar to a honeybee, and its close relation is the mason bee.

Above The tawny-mining bee (*Andrena armata*) is named after its abdomen and its mining activities in the ground. Hatching in its subterranean cell during the winter, the bee comes to the surface and flies off in the spring. Some members of the Andrenidae family live in colonies or "villages" which may number a thousand or more nests.

Left The carpenter bee (*Xylocopa violacea*) is very common in southern Europe and visits many kinds of flowers, especially those of wisteria. They provision holes in buildings and walls with pollen and lay eggs on top of it. Carpenter bees will sometimes become drunk on fermented nectar and fall to the ground completely exhausted.

What you can do

Training bees and wasps

Learning by association

1 On an outdoor table in summer, lay out a large sheet of white paper and, using a black marker, draw a number of bold shapes or patterns on them; a star, square, triangle, circle, X, radiating lines, and so on, but only one of each. Cover this with a sheet of clear plastic to protect it against rain or dew.

2 Place a small cup or bowl by each shape, and fill all but one with water. Fill the last one with a 25 per cent sugar solution, made by dissolving one part sugar in three parts boiling water and allowing it to cool. Refill the bowls as needed for about two weeks, but always keep the sugar solution by the same shape. If the sugar water is by, say, the triangle, and at the end of two weeks you move it to another location, you will find that most of the visiting bees and wasps will still tend to congregate at the triangle.

A sense of time

1 Place the sugar solution in the same location at exactly the same time every day for a one hour interval. Do this for two or three weeks, and you will find that for some time afterwards the bees will still show up at the appointed time, even if you do not bring the food.

2 You can even verify that they are the same individuals by placing dabs of bright paint or nail polish on the back of each bee's thorax as she drinks, taking care not to get any on her wings.

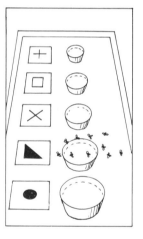

MINOR INSECT ORDERS

SELECTED ORDERS

The orders of Class Insecta covered have been the "Big Eight," which, combined, include about seven of every eight insect species. The remaining 16 or so orders in this class, though they include some rather significant and familiar families, are unceremoniously known as "minor" insect orders. In keeping with the practical nature of this book, most of it has been devoted to those orders you are most likely to encounter. However, there are no insignificant organisms, and though it is not intended as a field guide, *The Practical Entomologist* would be incomplete without at least mentioning a few of the smaller orders. Orders of note that are not described here include bristletails, springtails, termites, earwigs, chewing lice, sucking lice, scorpionflies, and fleas.

Top Differences in two castes of the termite (*Macrotermes gilvus*) are shown here; compare the large soldier with its powerful jaws and enlarged head, with the smaller worker.

Center A pair of common earwigs (*Forficula auricularia*) can be identified by the shape of their forceps, for those of the male are strongly curved, while those of the female are not. Wings are present, but rarely used.

Right Only a few millimeters long, the water springtail (*Podura aquatica*) relies on using surface tension for moving about.

Mayflies – Order Ephemeroptera

The order name means "to live but a day," which refers to the lifespan of its adults. Though some species may persist as adults for several days, most will die within 24 hours of molting into their winged form. These delicate insects are ill-equipped in the final stage of their life cycle to do anything but engage in frenzied mating swarms over the waters in which they lived as naiads. Large numbers of adults typically mature at once in a synchronized "hatch." Males seize any female that ventures into their swarm, copulate in flight, and die immediately afterward. Females enjoy a slightly longer life, submerging once again and laying eggs attached to underwater objects by short filaments. This happens within an hour or so of mating, and the females expire shortly afterwards.

Adult mayflies are yellowish or brownish, with an upward tilt to their long, slender abdomens. Most noticeable are the two or three very long, filamentous appendages extending from the tip of the abdomen. Also conspicuous are their many-veined, triangular forewings and roundish hind wings, which are held vertically over the back at rest. Mayflies, despite their name, are commonly seen in both spring and summer, and their presence is an indicator of a relatively unpolluted body of water. There are three families in this order: Ephemeridae, Heptageniidae, and Baetidae.

Cockroaches – Order Blattodea

The scourge of homeowners everywhere, cockroaches are not quite as bad as their reputation. While they may contaminate improperly stored food with their feces and produce disagreeable odors, they do not transmit human diseases. They forage at night and hide by day, and though winged, they usually escape by running. Intense campaigns to eradicate them have produced generations of survivors immune to most pesticides. Through such resilience, they have become one of the oldest types of winged insect, dating back more than 350 million years.

Cockroaches are brown, with bodies up to 60mm long that are flattened, helping them slip into narrow cracks to hide. Their pronotum extends far forward to produce a hooded appearance, and they have long, swept-back, thread-like antennae that help them to feel their way in the dark. Most species prefer warm, damp, subtropical

Top The naiad of the mayfly (*Ephemera danica*) crawls from the water and secures itself on a stem prior to the adult emerging. The aquatic immature stage may last three years in some species, but the ephemeral nature of the adult means that it is usually dead within a day.

Above Cockroaches are opportunists and scavengers that will eat all sorts of vegetable remains. Here Australian cockroaches (*Periplaneta australasiae*) devour an apple. Being very alert they scurry off when disturbed and have sharp spines on their legs to resist attack.

conditions, but representatives occur almost everywhere there is human habitation. Once considered part of the Order Orthoptera, they are now a separate order consisting of Families Blattidae and Blattellidae.

Mantids – Order Mantodea

Mantids are easily recognized by their long (up to 150mm), slender bodies and slow but graceful movements. They have two pairs of walking legs, and toothed, muscular forelegs that are well adapted for seizing prey with amazing swiftness. Typically held close to the body in a praying position, this front pair of legs has inspired the common name "praying mantis." Mantids employ detailed mimicry of green plant stems and leaves to stalk or ambush their victims.

The head has an amazing range of motion, so much so that the insect projects a quite human image as it looks over its shoulder or cocks its head to get a better view. Powerful mouthparts enable mantids to cut through the tough armor of most insects, their primary prey. Because of their predatory habits and voracious appetites, they are extremely beneficial, especially to gardeners. Females lay hundreds of eggs in a foamy mass that hardens into a tough, insulating case. Egg cases are usually attached to branches or man-made structures, but the mantids themselves can be found in dense vegetation. These, too, were once considered part of the Order Orthoptera.

Walkingsticks & Timemas – Order Phasmatodea

Masters of mimicry, walkingsticks and timemas are nearly invisible in their natural habitats. Walkingsticks, brown and up to 150mm long, are bizarrely modified to resemble leafless twigs. They even adopt a slow, swaying stride that copies the motion of wind-blown vegetation. Timemas resemble the leaves themselves. Unlike mantids, however, the camouflage of walkingsticks and timemas is purely defensive. Their metamorphosis is incomplete, and they can reproduce by parthenogenesis. The two families are Phasmidae and Timemidae.

Stoneflies – Order Plecoptera

The order name, which means "folded wings," refers to the broader hind pair, which is folded fan-like and covers most or all of the abdomen of individuals. The bodies, which are between 6mm and 64mm in length, are flattened, mirroring the streamlined nature of the naiads, which inhabit swiftly-flowing streams. Fairly long antennae and a pair of usually long cerci on the tip of the abdomen are also hallmarks of this order.

Another indicator of relatively pure water, stonefly-naiads are predators of well-aerated streams and rivers. Their brief adulthood is spent crawling around rocks near the water, hence the common name. Adults are capable only of a weak, fluttering flight. Families include Nemouridae, Capniidae, Taeniopterygidae, Peltoperlidae, Leuctridae, Pteronarcyidae, Perlodidae, Perlidae, and Chloroperlidae.

Dobsonflies, Alderflies, & Fishflies – Order Megaloptera

Because of their densely-veined wings, these insects were once classified in the Order Neuroptera, the net-veined insects, and even now the chief difference between the two orders is a matter of wing venation. Megaloptera, which means "ample wings," refers to the large hind wings. At rest, these are folded, roof-like, over the abdomen. Despite the large wings, adults are weak fliers, and are often seen fluttering near lights. Adults are between 9mm and 70mm in length, with long, swept-back antennae. They have chewing mouth-

Left Praying mantids are masters of disguise; often leaf-shaped, they may be colored green, brown or pink, sometimes varying as they grow older. They often lie by flowers waiting for visiting insects to arrive, whereupon they strike out with their battery of spines and impale the insect on their modified first pair of legs. They then start to eat the prey, dead or alive. This is a female praying mantis (*Galinthias amoena*), from Kenya, eating a fly.

Top The giant ant-lion (*Palpares weelei* of the Myrmeleontidae family), from the tropical dry forest of Madagascar, has large, spatula-shaped wings and relatively short antennae.

Right Ant-lion larvae live in groups in sandy soil where they each lie in wait at the base of a pit, waiting for unsuspecting prey to fall in. The sides of the sand pit are relatively steep, and any insect that attempts to cross the sand will fall directly into their waiting jaws.

Above This fabulous day-flying member of the lacewing family (Neuroptera) is found in south-western Turkey and is called *Neuroptera sinuata*. It has evolved colors and patterns similar to those of moths or butterflies.

parts, and the mandibles of male dobsonflies are massive and tusk-like, about three times longer than the head. Larvae are voracious aquatic predators with an appetite for black fly larvae. Metamorphosis is complete, with the larvae leaving their ponds or streams to pupate. Families include Corydalidae and Sialidae.

Net-veined Insects – Order Neuroptera

The order name means "nerve wings," a reference to the many-branched veins in their wings, which are all about equal in size and shape and are held roof-like over the abdomen when at rest. Like the Megaloptera, they have a weak, fluttering flight. Most adults are slender and delicate, measuring from 6mm to 45mm in length, with large compound eyes and long, slender antennae. Metamorphosis is complete, and the larvae of most species are very beneficial, preying on many destructive insects, such as aphids. Families include Coniopterygidae, Chrysopidae, Hemerobiidae, Sisyridae, Mantispidae, Myrmeleontidae, Ascalaphidae, Polystoechotidae, Dilaridae, and Ithonidae.

Caddisflies – Order Trichoptera

Caddisflies are notable for the aquatic larvae's habit, many species possess, of building portable protective cases around themselves by cementing together bits of sand or debris. Adults are quite moth-like, but lack the wing scales and coiled proboscis of an adult lepidopteran, having instead chewing mouthparts and fine hair on their wings. At rest, the wings are held roof-like over the abdomen. Most species are nocturnal, often fluttering around lights and hiding during daylight hours. Adults are between 7mm and 25mm long and gray or brown in color.

Above While some booklice are found on books, to which they are attracted by the vegetable content of fibers and glues, other booklice live in the open among leaf litter and debris. This winged psocid (*Psococerastis gibbosa*) has a marvellous violet sheen. Many other species of psocid, particularly those found in libraries, are wingless.

Right Caddis flies have aquatic larvae which are easy to find on the undersides of rocks. The larvae spin a fine web with their own silk and use this to filter food debris from the water of the stream. They also use tiny stones in the construction of their caddis cases, and the design and shape of a case can be a help in identifying the species.

GLOSSARY

abdomen: the posterior one of the three major subdivisions of an insect's body.

anal: toward the posterior end or side.

antennae: paired sensory appendages, one located on each side of an insect's head.

anterior: toward the head.

asexual reproduction: the production of viable offspring without the fertilization of eggs by sperm.

calypter: a lobe on either the posterior wing margin or close to the base of the wing on the thorax of certain fly families.

carrion: decaying flesh.

caste: a specialized segment of an insect colony, with very distinct functions.

cell: an area of wing membrane partially or completely bounded by veins.

cephalothorax: the first subdivision of a spider's body, composed of the head and thorax.

cerci: paired sensory appendages located at the posterior of the abdomen on some insects.

chitin: a component of the exoskeleton that is very tough, flexible, and has some degree of resistance to most chemicals.

chrysalis: a butterfly pupa not encased in a cocoon.

clubbed antennae: those that are abruptly enlarged at the tip.

cocoon: composed of silk fibers secreted by a

Cockchafer beetle (*Melolontha melolontha*) showing flabellate **antennae**.

This oil beetle (*Lydus syriacus*) clearly illustrates the three body parts: **head, thorax** and **abdomen**.

larva, this is the protective case in which the larva will pupate.

colony: a congregation of a single species of insects, living together cooperatively.

compound eyes: a pair of visual organs, one on each side of an insect's head, each composed of few to several thousand photoreceptive units radiating outward and terminating in a lens, all of which are joined to form facets of the eye.

coxa: the basal segment of an insect's leg.

cuticle: the non-cellular, secreted outer covering of an insect's body.

dimorphism: the occurence of two distinct forms of the same species.

dorsal: pertaining to the back of an organism, usually the upper surface.

elbowed antennae: those bent at a 90 degree angle near the middle.

elytrum(a): the hardened or leathery forewing(s) of beetles.

epidermis: the layer of living cells that underlie and secrete the insect's cuticle.

exoskeleton: an external, waterproof, protective body covering, composed of chitin, sclerotin, and waxes, which houses and supports the internal organs, muscles, and other tissues.

eyespots: protective markings on the wings of certain insects, intended to mimic large vertebrate eyes.

facet: the external surface of an individual unit of a compound eye.

femur: the third segment, counting from the body, of an insect's leg, found between the tibia and trochanter.

haltere: the vestigial hind wing of a fly, it functions as a stabilizer in flight.

hemelytra: distinctly thickened or leathery fore-wings of true bugs, which have overlapping membranous tips.

hemimetabolous development: incomplete metamorphosis, in which immature nymphs resemble adults of the species as soon as they hatch from the eggs. Their size increases with successive molts, as do the development of wings and sexual organs. Aquatic hemimetabolous insects in their immature form are called naiads.

holometabolous development: complete metamorphosis, in which the insect passes through four distinct stages in its life cycle (egg, larva, pupa, and adult), and the immature stages do not resemble the adult.

honeydew: a sweet excretion of many true bugs, especially aphids.

instar: a stage of insect development between molts. The first instar is the stage between hatching from the egg and the first molt.

invertebrate: an animal without a backbone.

labium: the most posterior of an insect's mouthparts, analogous to a lower lip.

labrum: the most anterior of an insect's mouthparts, analogous to an upper lip.

larva: the second stage in a holometabolous insect's life cycle, between the egg and pupa.

mandibles: a pair of insect mouthparts, analogous to jaws and normally used for chewing.

margin: an edge of an insect's wing.

maxillae: a pair of insect mouthparts, usually posterior to the mandibles and used for food

Lycus constrictus beetle showing the hardened, leathery **elytra**.

Pupa of a large white butterfly (*Pieris brassicae*).

handling.

mesonotum: the dorsal surface of the mesothorax.

mesosternum: the ventral surface of the mesothorax.

mesothorax: the middle segment of an insect's thorax, which bears the middle pair of legs and the forewings, if present.

metamorphosis: the transformation of a larva, nymph, or naiad into an adult insect.

metanotum: the dorsal surface of the metathorax.

metasternum: the ventral surface of the metathorax.

metathorax: the third, or most posterior, segment of an insect's thorax, which bears the third pair of legs and the hind wings if present.

molt: the shedding of the rigid cuticle to allow for growth or metamorphosis.

naiad: the immature form of an aquatic hemimetabolous insect.

notum: the dorsal surface of the thorax.

nymph: the immature form of a terrestrial hemimetabolous insect.

ocelli: simple eyes with a single lens that can detect varying degrees of light intensity but most probably cannot discern images.

ovipositor: the egg-laying appendage located at the posterior end of female insects.

palpi: paired sensory appendages of mouthparts.

parthenogenesis: asexual reproduction, in which eggs develop without being fertilized by sperm.

pedicel: the slender stalk of the abdomen found

153

on most members of the Order Hymenoptera.

pheromone: chemical messages emitted by insects.

pleuron: the side of an insect's abdomen.

posterior: away from the head.

proboscis: elongated, tube-like mouthparts, adapted for consuming liquids.

prolegs: soft, unjointed, paired appendages used for support and typically found on caterpillars.

pronotum: the dorsal surface of the prothorax.

prosternum: the ventral surface of the prothorax.

prothorax: the first, or most anterior, segment of the thorax, to which the first, or most anterior, pair of legs are attached.

pupa: the third, inactive stage of a holometabolous insect's life cycle, between the larval and adult stages, in which the insect transforms into an adult.

pupate: the process of becoming a pupa.

raptorial legs: thick, muscular forelegs of predatory insects, used to grasp and hold their prey.

sclerite: one of the hardened plates that compose the exoskeleton.

scutellum: a triangular sclerite of the mesonotum, located between the forewings and especially prominent among beetles and true bugs.

segment: a section of an insect's body or appendage, bordered by flexible membranous regions.

seta(e): hair-like sensory structure(s) that may be adapted to detect touch, smell, taste, or sound.

social insects: those species that live cooperatively in colonies and exhibit a division of labor among distinct castes.

Spotted longhorn beetle (*Strangalia maculata*) clearly showing the various **leg segments**.

The **scutellum** can easily be identified on these squashbugs (*Sagotylus confluentus*).

solitary insects: non-social insects that do not form long-term associations with other members of their species.

spiracle: a breathing hole, usually on the abdominal surface, through which gases enter and leave an insect's respiratory system.

sternum: the ventral surface of an insect's thorax and abdomen.

tarsus: the terminal section of an insect's leg, composed of one to five segments and attached to the tibia.

tergum: the dorsal surface of an insect's abdomen.

thorax: the middle of the three major subdivisions of an insect body, between the head and abdomen, to which all legs and wings are attached.

tibia: the fourth segment of an insect's leg, between the femur and tarsus.

trachea: part of the respiratory system, a tube through which gases move to and from internal organs.

trochanter: the second segment of an insect's leg, between the coxa and femur.

tympanum: an auditory organ consisting of a vibration-sensitive membrane on the abdomen or forelegs of grasshoppers, cicadas, and some moths.

ventral: pertaining to the side of the body opposite the back, usually the lower surface.

CONTACTS

Biological Equipment and Supplies

American Biological Supply Co.
1330 Dillon Heights Ave.
Baltimore, MD 21228
(301) 747–1797

Carolina Biological Supply Co.
2700 York Road
Burlington, NC 27215
(800) 547–1733

Connecticut Valley Biological Supply Co.
82 Valley Road
South Hampton, MA 01073
(413) 527–4030

Edmund Scientific
101 East Gloucester Pike
Barrington, NJ 08007
(609) 573–6250

Fisher Scientific
52 Fadem Road
Springfield, NJ 07081
(800) 441–2374

Frey Scientific Company
905 Hickory Lane
Mansfield, OH 44905
(216) 589–1900

Nova Scientific Corporation
111 Tucker St.
P.O. Box 500
Burlington, NC 27215
(919) 584–0381

Ward's Natural Science Establishment
5100 West Henrietta Road
Box 92912
Rochester, NY 14692
(800) 962–2660

Watkins and Doncaster
Conghurst Lane
Four Throws
Hawkhurst
Kent TN18 5ED, UK
05805–3133

Worldwide Butterflies Ltd
Compton House
Nr. Sherbourne
Dorset DT4 QN, UK
0935–74608

Entomological Organizations

Butterfly World
Tradewinds Park
3600 West Sample Rd.
Coconut Creek, FL 33073
(305) 977–4400

Calloway Gardens Day Butterfly Center
Pine Mountain, GA 31822
(800) 282–8181

Hole-in-Hand Butterfly Farm
Hazleton, PA 18201

Lepidoptera Research Foundation
Santa Barbara Museum of Natural History
2559 Puesta del Sol Road
Santa Barbara, CA 93105

Lepidopterists' Society
Department of Biology
University of Louisville
Louisville, KY 40208

Xerces Society
Department of Zoology and Physiology
University of Wyoming
Laramie, WY 82071

The Amateur Entomological Society
Publications Agent
137 Gleneldon Road
Streatham, London SW 16, UK

Amateur Entomologist
109 Waveney Drive
Springfield
Chelmsford
Essex CM1 5QA, UK

E.W. Classey Ltd
Natural History Booksellers
P.O. Box 93
Faringdon
Oxon, UK

The Royal Entomological Society of London
41 Queen's Gate
London
SW7 5HU, UK
071–5848361

INDEX

ACKNOWLEDGMENTS

Quarto would like to thank the following people and organisations for permission to reproduce copyright material.

Key: bl = bottom left; br = bottom right; tr = top right; tl = top left; l = left; r = right; t = top; c = center; b = bottom

Premaphotos Wildlife
(c) **K.G. Preston-Mafham** 2; 4; 5tl, tr, cl, cr & br; 6; 9br; 11t & b; 19; 20t & b; 23; 26; 27; 31; 35t; 36; 41l; 53; 69; 70b; 71t & b; 72; 76l & r; 78l & r; 79; 81b; 83t, c & b; 84t; 89tl, tr & b; 90l & r; 95l; 97; 98r; 99; 105t & br; 107; 111t & b; 114b; 116tr; 118; 119t; 130r; 132; 134l; 135; 138b; 139t; 141l & r; 144l; 147t & b; 148; 149t & b, 152–3 (c) **R.A. Preston-Mafham** 38; 85; 98l; 101t & b; 116tl; 133b; 144r; 145 (c) **J. Preston-Mafham** 81tr

Wildlife Matters: John Feltwell
5bl; 10t & b; 22; 32; 37; 41r; 43; 46; 51; 52; 55; 58; 59; 61b; 62; 63; 64; 65; 75l & t; 86; 100t & b; 102; 113l & r; 114t; 115; 119b; 121; 123; 140t
Wildlife Matters: Michael Tweedie
7; 9t & bl; 25; 28; 33; 34; 35b; 39; 40; 50; 70t; 73; 81tl; 84b; 91t & b; 92t & b; 93; 95r; 96t & b; 100c; 103; 105bl; 109; 116b; 120; 123t, c & b; 124; 125; 128l & r; 129; 130l; 133t; 134r; 137; 138t; 139b; 140b; 146t, c & b; 150; 151t & b
Wildlife Matters: British Museum (Natural History)
21, 61t
Wildlife Matters: taken while on Project Wallace, courtesy of the Royal Entomological Society of London
49